Advances in Mathematics: Theory, Methods & Applications

RIVER PUBLISHERS SERIES IN MATHEMATICAL, STATISTICAL AND COMPUTATIONAL MODELLING FOR ENGINEERING

Series Editor:

MANGEY RAM
Graphic Era University, India

Applied mathematical techniques along with statistical and computational data analysis has become vital skills across the physical sciences. The purpose of this book series is to present novel applications of numerical and computational modelling and data analysis across the applied sciences. We encourage applied mathematicians, statisticians, data scientists and computing engineers working in a comprehensive range of research fields to showcase different techniques and skills, such as differential equations, finite element method, algorithms, discrete mathematics, numerical simulation, machine learning, probability and statistics, fuzzy theory, etc.

Books published in the series include professional research monographs, edited volumes, conference proceedings, handbooks and textbooks, which provide new insights for researchers, specialists in industry, and graduate students.

Topics included in this series are as follows:

- Discrete mathematics and computation
- Fault diagnosis and fault tolerance
- Finite element method (FEM) modeling/simulation
- Fuzzy and possibility theory
- Fuzzy logic and neuro-fuzzy systems for relevant engineering applications
- Game Theory
- Mathematical concepts and applications
- Modelling in engineering applications
- Numerical simulations
- Optimization and algorithms
- Queueing systems
- Resilience
- Stochastic modelling and statistical inference
- Stochastic Processes
- Structural Mechanics
- Theoretical and applied mechanics

For a list of other books in this series, visit www.riverpublishers.com

Advances in Mathematics: Theory, Methods & Applications

Editors

Akshay Kumar

Graphic Era Hill University, India

Mangey Ram

Graphic Era Deemed to be University, India

River Publishers

Routledge
Taylor & Francis Group
NEW YORK AND LONDON

Published 2025 by River Publishers
River Publishers
Broagervej 10, 9260 Gistrup, Denmark
www.riverpublishers.com

Distributed exclusively by Routledge
605 Third Avenue, New York, NY 10017, USA
4 Park Square, Milton Park, Abingdon, Oxon OX14 4RN

Advances in Mathematics: Theory, Methods & Applications / by Akshay Kumar, Mangey Ram.

Routledge is an imprint of the Taylor & Francis Group, an informa business

ISBN 978-87-7004-669-5 (hardback)
ISBN 978-87-7004-823-1 (paperback)
ISBN 978-87-7004-671-8 (online)
ISBN 978-87-7004-670-1 (master ebook)

Contents

Preface

Mathematics plays a vital role in the branch of physical science and applied mathematics, mathematical modeling, and engineering. Novel ideas and applications are used to present concise and clear introductions to engineering mathematics, which move from the fundamentals to the complex mathematical notions required for all engineering specializations. The main aim of this book is to guide readers through the core concepts of mathematics and help them understand how those concepts apply in engineering science. The contents of this book address profitable and effective interaction between engineers and mathematicians, as well as the cross-fertilization and collaboration that result when mathematics is engaged with the advancement of science and technology.

This book covers the following topics: numerical methods, integral equations, differential equations, engineering mathematics, technical analysis, cost analysis, system and network configuration, advanced-level mathematics, approximation theory, and optimization.

Acknowledgement

The editors acknowledge River Publishers for this opportunity and for the excellent support provided to us to complete this book. Also, we would like to thank all the chapter authors and reviewers for their availability for this work.

Akshay Kumar
Graphic Era Hill University, India

Mangey Ram
Graphic Era Deemed to be University, India

List of Figures

List of Tables

List of Contributors

Arora, G., *Department of Mathematics, School of Physical Sciences and Chemical Engineering, Lovely Professional University, India; E-mail: geetadma@gmail.com*

Bhadula, Rakesh Chandra, *Department of Mathematics, Graphic Era Hill University, India; E-mail: Rakeshbhadula.14@gmail.com*

Dubey, Ramu, *Ramu, Department of Mathematics, J. C. Bose University of Science and Technology, YMCA, India; E-mail: rdubeyjiya@gmail.com*

Fatima, Aman, *Department of Mathematics, Jaypee Institute of Information Technology, India; E-mail: amanfatima204@gmail.com*

Gupta, Gourav, *Department of Mathematics, Lovely Professional University, India; E-mail: gouravgupta333@gmail.com*

Gupta, J., *GNA University, School of Natural Sciences, Phagwara, Punjab, India; E-mail: jayagupta9295@gmail.com*

Gupta, Pooja, *Department of Mathematics, J. C. Bose University of Science and Technology, YMCA, India; E-mail: poojaguptaiitr@gmail.com*

Hooda, D. S., *Kurukshetra University, India; E-mail: ds_hooda@rediffmail.com*

Jangid, Riddhi, *School of Computational and Integrative Sciences, Jawaharlal Nehru University, India; E-mail: riddhi40_sit@jnu.ac.in*

Kala, V. N., *G.B. Pant Engineering College, India; E-mail: vnkala67@gmail.com*

Kaur, Lakhveer, *Department of Mathematics, Jaypee Institute of Information Technology, India; E-mail: lakhveer712@gmail.com*

Ketana, K., *Department of Mathematics, BITS Pilani Hyderabad Campus, India; E-mail: p20210113@hyderabad.bits-pilani.ac.in*

Kumar, Ankit, *Department of Mathematics, Graphic Era Hill University, India; E-mail: ankitramkumar620@gmail.com*

Kumar, Manoj, *Department of Mathematics, Graphic Era Hill University, India; E-mail: docmanojkumar64@gmail.com*

Lazes, Chosching, *School of Computational and Integrative Sciences, Jawaharlal Nehru University, India; E-mail clazes1297@gmail.com*

Mishra, S., *Department of Mathematics, School of Physical Sciences and Chemical Engineering, Lovely Professional University, India; E-mail: mishrasam73@gmail.com*

Mudgal, T. R., *Department of Applied Sciences & Humanities, State Institute of Engineering & Technology, India; E-mail: trmudgal@gmail.com*

Murugesan, K., *Department of Mathematics, National Institute of Technology, India; E-mail: murugu@nitt.edu*

Radhika, T. S. L., *Department of Mathematics, BITS Pilani Hyderabad Campus, India; E-mail: radhikatsl@hyderabad.bits-pilani.ac.in*

Ram, Neetu, *Department of Mathematics, Harsh Vidhya Mandir PG Collage Raisi, India; E-mail: dr.n.chaudhary.rke@gmail.com*

Rani, Deepika, *Department of Mathematics, Dr. B. R. Ambedkar National Institute of Technology, India; E-mail: ranid@nitj.ac.in*

Rani, Richa, *Department of Mathematics, Guru Gorakhnath Ji Govt. College Hisar, Haryana, India; E-mail: richasardana7@gmail.com*

Singh, Gajendra Pratap, *School of Computational and Integrative Sciences, Jawaharlal Nehru University, India; E-mail: gajendra@jnu.ac.in*

Singh, Jaipal, *D.B.S.(PG) College, India; E-mail: jaipalsingh.dbs@gmail.com*

Singh, Kuldeep, *Department of Mathematics, J. C. Bose University of Science and Technology, YMCA, India; E-mail: kuldeepsinghbhati599@gmail.com*

Singh, Sukhveer, *Department of Mathematics, Graphic Era University, India; E-mail: sukhveer.singh@thapar.edu*

Singh, Teekam, *Department of Computer Science and Engineering, Graphic Era (Deemed to be) University, India; E-mail: teekamsingh.iitr@gmail.com*

Singla, Jajji, *Chitkara University Institute of Engineering and Technology, Chitkara University, India; E-mail: jajjisingla.js@gmail.com*

Trokhimchuck, Petro, *Lesya Ukrainka Volyn National University, Ukraine; E-mail: Trokhimchuck.Petro@vnu.edu.ua*

Ujarari, Chandan Singh, *Department of Mathematics, Graphic Era Hill University, India; E-mail: ujarari0000@gmail.com*

Verma, Pallavi, *Department of Mathematics, Jaypee Institute of Information Technology, India; E-mail: pallavi4verma@gmail.com*

Yadav, Vandana, *Department of Mathematics & Scientific Computing, National Institute of Technology, India; E-mail: vandanarao1229@gmail.com*

List of Abbreviations

ADC	Apparent diffusion coefficient
BE	Burgers equation
BVP	Boundary value problem
CAD	Computer-aided design
CAM	Computer-aided manufacturing
CF	Certainty factor
CFD	Computational fluid dynamics
COPTEM	Crude oil pipeline transportation emissions model
CPN	Colored Petri net
CSS	Continuum surface stress
DEA	Data envelopment analysis
FDM	Finite difference method
FPN	Fuzzy Petri net
GFPN	Generalized fuzzy Petri net
HLPN	High-level Petri net
IBVP	Initial boundary value problem
IVP	Initial value problem
MAE	Mean absolute error
ML	Machine learning
MMAE	Method of matched asymptomatic expansions
NLPDE	Nonlinear partial differential equation
NMR	Nuclear magnetic resonance
ODE	Ordinary differential equation
PA	Polymetric analysis
PCE	Polynomial chaos expansion
PDE	Partial differential equation
PN	Petri net
PSO	Particle swarm optimization
SPN	Stochastic Petri net
S-S	Steady state

SVR	Support vector regressor
VAM	Vogel approximation method
VOF	Volume of fluid
WKB	Wentzel-Kramers-Brillouin

List of Notations

Q_1	Inlet discharge
Q_2	Outlet discharge
A_1,	Cross-sectional areas of the inlet discharge
A_2	Cross-sectional areas of the outlet discharge
h_f	Head losses due to fractions
R_e	Reynolds number
H_1'	Static head just upstream from the location of the leakage
$L - l_i$	Downstream from the i th hole
K_r	Resistance coefficient
F	Leak rate
k	Discharge coefficient

1

Natural Concept of the Foundations of Mathematics

Petro Trokhimchuck

Lesya Ukrainka Volyn National University, Ukraine
E-mail: Trokhimchuck.Petro@vnu.edu.ua

Abstract

The necessity of creation of a natural concept of the foundations of mathematics is discussed. A short historical analysis of the development of modern mathematics with this point of view is presented. It is shown that the foundations of mathematics should be closely related to the nature of mathematics. Since the development of modern science has led to the expansion of both mathematics itself and the areas of its applications, the need for a universal approach has arisen, which allows us to describe all these theories from a unified point of view. Six rules that such theories must satisfy are formulated. Polymetric analysis is given as an example of such a theory, which can be presented as a natural concept of the foundations of mathematics.

Keywords: Foundations of mathematics, natural concept, Pythagor, Euclides, Newton, Euler, Wolf, polymetrical analysis

1.1 Introduction

The problem of the foundations of mathematics should be considered in the context of the formation and development of human civilization and culture [1–25]. That is why a brief historical overview of this problem is given.

It is obvious that the basics of mathematics should be approached taking into account the sayings of R. Bacon and R. Descartes: "Science is as much science as there is mathematics in it" [16].

It should be noted that, according to V. Arnold [6], mathematics is precise knowledge, science is ordered knowledge, and, in addition, there is also unordered knowledge [8]. Such a classification allows for a more accurate assessment of the role of mathematics in modern science. The foundations of mathematics can be considered and analyzed from two points of view [8]. The first is structural, based on one or more sections of mathematics, which can be used to obtain its other sections. Examples of such an approach are logical, formal, and intuitionistic conceptions of the foundations of mathematics [17]. As K. Gödel's research showed, these systems are incomplete [15]. In other words, you cannot build a theory of the foundations of mathematics based on only one of its chapters. That is why one of the creators of the logical concept of the foundations of mathematics, A. Whitehead, abandoned his views on the concept in favor of the organismic concept, that is, in the direction of intuitionism [18].

The second is natural, based on the nature of mathematics. The nature (subject) of mathematics is the analysis, synthesis, and formalization of any field of knowledge or activity. However, it is nature of each science. Therefore, this concept shows the bond between mathematics and other sciences. In this sense, mathematics corresponds to the phrase "Mathematics is the queen of sciences" [7, 8]. Roughly speaking, the theory of the foundations of mathematics must include the procedure of creating a corresponding theory and must be an open system.

An example of such an approach is polymetric analysis [8, 16]. If the structural approach was good for the development of certain chapters of mathematics, in particular mathematical logic and logical types of B. Russell, then the natural approach corresponds to the current state of mathematics in connection with the development of computer sciences. The main thesis of computer science, according to A. Ershov, is the formalization of the thesis of the Canadian philosopher L. Hall: "Everything that comes from the head is intelligent" [2].

The formalization of this thesis is closely connected with the theory of open systems and systems with a variable hierarchy [16]. Based on a detailed analysis of such theories, six criteria that such theories must satisfy were developing and given [8].

1.2 Main Concepts of the Foundation of Mathematics

In general, mathematics as a science arose at the birth of the first civilizations [8, 16]. In general, systems of knowledge of these civilizations may be represented as two types: open and closed. These terms refer to which ritual or religious system was primary. Thus, Ancient Sumer belongs to open-type civilizations. Iatromathematics, astronomy and astrology, spherical geometry, etc., were created there. The civilizations of ancient Egypt belong to the closed type. Algebraic methods were developed there along with geometrical methods, and esotericism played a significant role. It was there that the optimal encrypted systems of knowledge known as the Tablet of God Thoth and the Heliopolis Ogdoada were obtained.

Civilizations of the open type included Phoenician, ancient Greek, Roman, etc. civilizations. The civilizations of Latin America (Incas, Aztecs, Toltecs, Equor, etc.) belonged to the civilizations of the closed (esoteric) type [16].

It should be noted that, to one degree or another, the elements of mathematics were inherent in the ancient Indian and Chinese civilizations. And where exactly do we thank India with the designation of modern numbers [8]?

There were attempts to combine closed and open knowledge systems into a single system. One of the most successful attempts should be considered the Pythagorean system, which based this unification on mathematics. The phrase "Numbers rule the world" belongs to Pythagoras. Somewhat later, Plato formulated a statement about three types of numbers: mathematical (roughly speaking, pure mathematics), sensible (in modern language, applied mathematics), and ideal (in modern language, various kinds of numerology).

Antiquity played an important role in the formation and development of modern science. It was here that Aristotle proposed the first classification of sciences. She is the basis of the modern classification of sciences [16].

It was in the era of antiquity that mathematics as an exact science was singled out as a separate branch of human culture and science. We owe this to Euclid and his "Elements" [11, 16]. It was his axiomatic method that played a significant role in the establishment of mathematics as a science and later in the development of other theoretical sciences.

In the third-century BC the Alexandrian Museum was created. It was created by the founder of the Ptolemaic dynasty on the initiative of Aristotle's student Demetrius Falersky [16]. Alexandria Museum was main scientific and cultural centers of the Hellenistic world and first scientific organization in modern sense of this word [16].

Scientists from many fields of science, invited from different countries of the Mediterranean, lived and worked in the Alexandria Museum. The Alexandria Museum was headed by priests of the highest rank, who were personally appointed by the Ptolemies, and then by the Roman emperors.

Scientists working in the museum were fully supported by the Ptolemies (later the Roman emperors). The muses reached their peak in the third century, and according to some scholars, up to the third–second centuries BC. Great successes were achieved in mathematics, astronomy, natural science, geography, medicine, as well as philology and grammar [16].

Alexandria Museum was destroyed in AD 272–273 in the time of the Roman emperor Aurelian. Later, it was the Alexandrian Mouseion that became an example for the creation of organizations of this type at the courts of the Baghdad and Cordoba caliphs, the rulers of Merage and other Muslim rulers, as well as the French, Prussian, Spanish, Russian, and other Academies of Sciences [16].

It should be noted that in the Middle Ages, science and mathematics developed more intensively in India, China, and Islamic countries. We owe the development of science in Islamic countries to Prophet Muhammad and his phrase "Muslim men and women should learn everything and from everyone, including the Chinese" [8, 16].

The Renaissance in Europe began with the translation from Arabic of the works of Euclid, Aristotle, Plato, Archimedes, etc. That is why the works of J. Bruno and Galileo were written in the form of dialogues based on the example of Plato. And R. Bacon wrote his main works: Opus Maius, Ohus Minus, and Opus Tertius based on the example of Euclidean "Elements" [16].

But the main contribution to the formation of modern mathematics and further formalization of knowledge, including the creation of theoretical sciences, was made by Rene Descartes [3, 16].

The main rules of the Descartes method are as follows [3, 16]:

Rule 1: Accept nothing as true. We must test our hypothesis on things we do or say. Are we doing it because everyone is doing it, or is it self-evident? This is not meant to critique everything, but it allows us to dissociate ourselves in the moment from being too quick to act without taking a moment to test our hypothesis and see it from different vantage points.

Rule 2: Divide into simple parts. When a problem arises, dissect it to its lowest indivisible number. Problems often surface something that is deeper than what is evident at the surface. I often tell coaches that the problem isn't the problem; we need to unravel by chunking through myriad information and discover the basic building blocks.

Rule 3: Move from simple to complex. Simplicity isn't simple-minded; it is complex to make something simple. Once we have the simple building blocks, we are looking at the multiplier effect of a system.

Too often, when leaders look at teams, they see them as one plus one, but the members we have in the team are not an addition but a multiplier effect. If we can utilize the requisite power of synergy, we are able to reduce wasting resources and can become precise in our execution.

Rule 4: Test the reasons. How did we arrive at this reasoning? Testing and re-testing allow us to conduct a simulation of whether our hypothesis is founded on grounds of accuracy or fallacy. The reasons will show the internal motivation and drive that fuel the outcome.

Thanks to this method, Descartes created analytical geometry – a synthesis of number theory and geometry and initiated the creation of theoretical mechanics, the creation of which was completed by I. Newton [1, 16].

In addition, Newton improved Descartes' method and formulated his four rules of inference in physics, which is nothing but the application and extension of the Euclidean method to any field of knowledge. Let us outline these rules [1]:

Rule 1: No more causes of natural things should be admitted than are both true and sufficient to explain their phenomena.

Rule 2: Therefore, as far as possible, we must attribute the same causes to the manifestations of the nature of the same species.

Rule 3: Those qualities of bodies that cannot be intended and remitted and that belong to all bodies on which experiments can be made, should be taken as qualities of all bodies universally.

Rule 4: In experimental philosophy, propositions gathered from phenomena by induction should be considered either exactly or very nearly true not withstanding any contrary hypotheses, until yet other phenomena make such propositions either more exact or liable to exception.

Roughly speaking, from the time of Descartes and Leibniz, theoretical sciences have developed in two ways. The first is based on the Descartian method [3] and the four Newtonian rules of conclusions in physics [1]. It is a further development of the Euclidean method. The second may be represented as the development of the Leibniz idea of searching the universal calculation with the help of formal logic [24].

The foundations of mathematics have two main directions of development: natural and structural. Natural is based on the nature of mathematics and, therefore, implies the creation of a method or theory common to all branches of knowledge and a structural one that puts in the foundations of mathematics [2, 17] any of its sections – logic (Russell and Whitehead), geometry (Hilbert), arithmetic (Hilbert and Bernays), or a complex of mathematical disciplines (intuitionism, Brouwer, Gaiting, Kleene [23], and Wolfram [22]).

Newton's main merit is that he not only used Euclid's axiomatic method to implement R. Bacon's thesis "Science is as much science as there is mathematics in it" but also formulated the basic rules of this axiomatization [16].

The Newtonian method was used by Newton himself in the creation of classical mechanics (completion of the synthesis of terrestrial and celestial mechanics initiated by R. Descartes into a single system [8]) [1]. This method was also used by B. Russell when creating his inductive theory of logical types [10]. The Newtonian method may be used for the creation of intuitionism theories too [17].

Ch. Wolf (1716) is defining mathematics in the next way [16]: "Mathematics is a science to measure everything that can be measured. Usually it is described as a science of quantities, the science of magnitudes, that is, of those things that can increase or decrease. Since all finite things can be measured in all that they have in themselves finite, that is what they are, there is nothing in the world that could not be applied to mathematics, and since one cannot have any more precise knowledge than when the properties of things it is possible to measure, then mathematics leads us to the most perfect knowledge of all possible things in the world."

An important role in the development and understanding of mathematics was played by the definition of a mathematical quantity according to Euler [8, 16]:

1. First of all, everything that can be added or something can be taken away is called a mathematical quantity.
2. There are a lot of different kinds of quantities that cannot be counted, and various branches of mathematics come from them, each of which deals with its own kind of quantities. Mathematics in general is the science of quantities, concerned with finding ways to measure them.
3. However, it is impossible to determine or measure one quantity rather than to take as known another quantity of the same kind and indicate the ratio in which it is to it.
4. We must make everywhere such complete lists and such general surveys as to be sure that nothing is omitted.

At the same time, E. B. de Condillac created the first general theory of systems [12], according to which "Every system is nothing but the arrangement of various parts of any art or science in a known order in which they are interconnected. Parts, which include the explanation of other parts are called principles, and the system is all the more perfect, the smaller the number of its principles: it is desirable even that their number should be reduced to one."

The second way is due to the development of the ideas of W. Leibniz about the creation of universal calculus based on formal logic [8, 16, 24]. This leads to the creation of mathematical logic, logical foundation of mathematics, and Kleene metamathematics [23]. Kleene added and developed mathematical logic by calculations [7]. Later they attributed this system to intuitionism [7, 17]. S. Wolfram metamathematics can also be attributed to intuitionism (constructivism) [22]. His metamathematics is a deductive extension of his computer mathematics.

Mathematical logic is one of basic chapters of theoretical computer science [2, 8, 16]. The way of development of mathematical logic is discussed next. W. Leibniz introduced a law of sufficient satisfaction [24]. He hoped to build a universal calculus with the help of logic.

Boole, Morgan, Peano, Russell, and Whitehead created mathematical logic [8, 10]. But only Russell and Whitehead developed his idea of creating a universal calculus. They believed that all mathematics can be built using mathematical logic. But since the limits of application and construction are much larger than the subject area of mathematical logic, paradoxes of the inter-hierarchical type arose, as pointed out by A. Poincaré and K. Gödel [15, 17].

The concept of completeness is basic in mathematical logic and was developed by K. Gödel [15].

Therefore, let us now consider the concept of completeness and its connection with the foundations of mathematics. In mathematical logic, a theory is called complete if any syntactically correct closed formula or its negation is provable in this theory [7, 10, 16].

Turing completeness is a characteristic of an executor (a set of computing elements) in the theory of computability, meaning the ability to implement any computable function on it. In other words, for each computable function, there is an element that calculates it (for example, a Turing machine) or a program for an executor, and all functions that are computed by a set of calculators are computable functions (perhaps with some encoding of input and output data) [7, 10, 16].

The most widely used programming languages are Turing-complete. This applies to both imperative languages such as Pascal and functional languages

(Haskell) and logic programming languages (Prolog). Some programming languages (Haskell, C++) are compile-time Turing-complete in addition to run-time Turing completeness [2, 16].

Turing-complete normal Markov algorithm, 2-tag system, rules of cellular automaton, inhibitory Petri net, lambda calculus with beta reduction and unrestricted grammars are examples of Turing-complete systems [10].

Examples of Turing-incomplete formalisms are finite automata, primitive recursive functions, context-free, and regular grammars [2, 16].

Kurt Gödel showed a whole class of first-order theories that are incomplete [15]. Moreover, the formula asserting the consistency of a theory is also non-derivable by means of the theory itself (see Gödel's incompleteness theorems). This conclusion was of great importance for mathematics, since formal arithmetic (and any theory containing it as a subtheory) is just such a first-order theory, and therefore incomplete.

His two incompleteness theorems are represented in [15] and [17].

From the system point of view, these theorems may be represented as particularly formalization of rule 3 from Newtonian rules of conclusions in physics.

The problem of completeness may be transited on all closed systems. Therefore, Gödel's theorems showed that in order to create a universal theory that could claim to be a theory of the foundations of mathematics, there must be an open system-type theory with a shifting hierarchy.

In order to circumvent the problem of incompleteness in mathematical logic, B. Russell created the theory of logical types [10]. Practically, this is the application of the fourth rule of inference in Newtonian physics to mathematical logic [7, 8]. As the formal system becomes more complicated, it should be supplemented with appropriate operations (predicates). The number of such types can be arbitrarily large. This can also be considered a retreat to intuitionism (constructivism).

D. Hilbert tried to put geometry in the foundations of mathematics [17] and arithmetic together with his student P. Bernays [17]. This path was called the formal approach in the foundations of mathematics. But even in this way, it was not possible to create a universal calculus according to Leibniz. The third approach in the foundations of mathematics, intuitionistic, was proposed by L. Brouwer [7, 17]. It is more pragmatic; its goal is to create an optimal formalized system for a specific problem or class of problems. At the same time, the number of formalization systems can be arbitrarily large.

P. Cohen showed that the axioms of choice do not depend on the logical axioms and thus supplemented K. Gödel's results on incompleteness based on similar considerations [8, 16]. And this means that in order to build an uncontroversial theory of the foundations of mathematics, it is necessary to return to the Euclidean–Descartian–Newtonian line.

Kleene created the recursive calculations, which are the development of Russell types, and practically realized the synthesis of the concept of Russell and Brouwer [23].

All these concepts represented the structural line of the foundations of mathematics because it is based on structure mathematics [7].

From the system point of view, the Newtonian line is fuller as Leibnizian [1, 7, 8]. Real computing operations are based on computer arithmetic. The further development of this direction the mathematics was made by S. Wolfram, the creator course of computer mathematics "Mathematics", who included other branches of mathematics in his metamathematics, including the theory of graphs [22], and made further development in this direction of foundations of mathematics. At the same time, this is much more an implementation of the Brouwer program than Leibniz–Russell–Whitehead. Therefore, we must search other systems, which can be metamathematics and metascience.

Thus, the program of foundation of mathematics must include basic elements of its nature: analysis, synthesis, and formalization [2, 7, 8, 16, 17]. Structural concepts of foundation of mathematics are not corresponding to this condition. Therefore, A. N. Whitehead abandoned from his logical concept and said that it must be an organism concept [18]. One of the ways to solve this problem should be a system approach, which is basing on a more detailed formalization of the Euclidean–Descartian–Newtonian concept [8].

As was shown, theories of this type must satisfy the following conditions [8, 16]:

1. It must be an open theory or a theory with variable hierarchy.
2. This theory must be having a minimal number of principles.
3. It must be based on the nature of mathematics (analysis, synthesis, and formalization all possible knowledge).
4. We must create a sign structure, which unites verbal and nonverbal knowledge (mathematical and other) in one system.
5. We must have a system, which is an expert system of the existing system of knowledge and may be used for the creation of new systems of knowledge.
6. The principle of continuity must be true for all science.

These conditions must be used for the creation of any dynamic science.

An example of a theory satisfying these six conditions is polymetric analysis [8].

1.3 Polymetric Analysis as a Natural Concept of the Foundations of Mathematics

The basic task of PA is the formalization of triple minimum in the process of formalization the knowledge: philosophical (methodological), mathematical, and concrete scientifically [16].

The main elements of this theory and their bonds with other sciences are represented in Figure 1.1 [16].

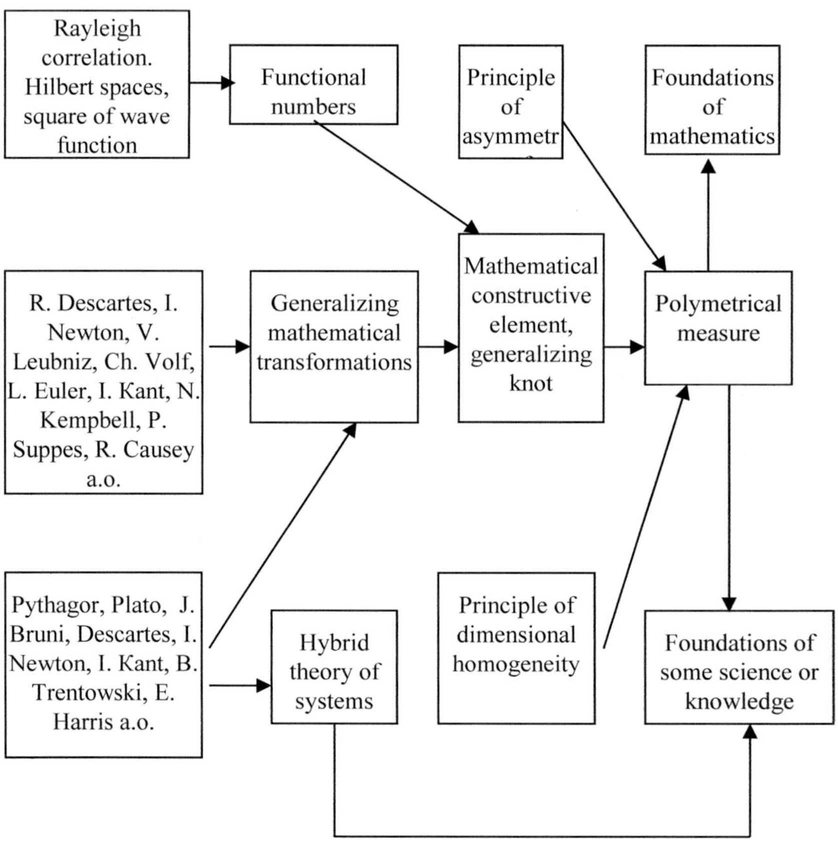

Figure 1.1 Scheme of polymetric method and its place in modern science [16].

The concept of polymetric analysis is based on the idea of a triple minimum: methodological, mathematical, and specifically scientific [6]. It should be noted that this idea was expressed by J. Bruno in 1590 in his work "The Triple Minimum in Measurement" [6]. In this work, he pointed out that there are three indivisible quantities: the number in mathematics, the atom in physics, and the monad in philosophy, from which measurements should begin [6]. Later, W. Leibniz united this single monad, which included the three elements of J. Bruno [6]. However, it was an "atomic" concept. In this sense, polymetric analysis is the result of a systematic synthesis of three main fields of knowledge and is a systemic concept.

The main elements of this synthesis are functional numbers, generalized mathematical transformations, and theory of hybrid systems.

Roughly speaking, we have to go back to Newton (his four rules of inference in physics) [1]. It was these rules, although they were written as a supplement to Newton's axiomatic method, that played a significant role in the formalization of knowledge and the creation of theoretical sciences. The logical direction itself was initiated by W. Leibniz who tried to use logic to build a universal calculus; although from a computational point of view, integral–differential calculus is more powerful than mathematical logic [5–7].

The main mathematical elements of PA are functional elements of quadratic forms (roughly speaking, generalizing multifunctional matrixes) [7]. This group is represented by functional parameters and functional numbers.

The second group of elements of PA is mathematical and other transformations, which allows connecting mathematics (roughly speaking, the ordering and formalization) with proper science [7]. Quantitative and qualitative transformations are introduced here, which correspond, roughly speaking, to primary and secondary measurements. Based on functional numbers and these transformations, the main elements of calculations – generalized constructive mathematical elements – were constructed [16].

The third group of elements of PA has system nature: corresponding theories and systems of analysis, synthesis, and formalization of various areas the knowledge [16]. Among them:

1. Polymetrical theory of measure and measurements. This theory is based on two principles: principle of asymmetry for primary measurements and principle of dimensional homogeneity for secondary measurements. For this case, the generalized mathematical element is the element of polymetrical measure, which includes the procedure of measurements.

2. Information lattice – the manifold of generalized mathematical elements. In this case, the generalized constructive mathematical element is the generalized knot of the informative lattice.
3. Theory of information calculation, which is constructing on the informative lattice.
4. Hybrid theory of systems – universal theory, which allows to classify all possible concepts and theories by degrees of their complexity. This theory is based on two criteria: criterion of reciprocity and criterion of simplicity and parameter of connectedness. The criterion of reciprocity is the principle of the creation of the corresponding mathematical constructive system (informative lattice). The criterion of simplicity is the principle of the optimization of this creation [7].

This concept allows formalizing the Pythagorean phrase "Numbers are ruling the World" [7].

Detailed main axiomatic of PA is represented in [16]. Now we represent only necessary elements of this axiomatic [16].

Definition 1. *Mathematical construction (constructive)* is the set of all possible elements, operations, and transformations for the resolution of the corresponding problem. The basic functional elements of this construction are called constructive elements [7].

For further describing the theory of information calculations (computing), we need to introduce the notion of information lattice [7].

Definition 2. A generalized knot or generalizing constructive mathematical element is a ratio

$$ {}^{stqo}_{nmab}B^{\alpha\beta\gamma\ldots}_{ijl\varphi} = A_{i_1}\ldots A_{i_\alpha}\overline{A}_{i_1}\ldots\overline{A}_{i_\alpha}O_{k_1}\ldots O_{k_\gamma}\overline{O}_{p_1}\ldots N_{\varphi_{ij}}, \qquad (1.1) $$

where $N_{\varphi_{ij}}$ – functional numbers, A_{i_1} – straight qualitative mathematical transformations, $\overline{A}_{i_1}$ – opposite qualitative mathematical transformations, O_{k_1} – straight quantitative mathematical transformations, $\overline{O}_{p_1}$ – opposite quantitative mathematical transformations [16].

The manifold of generalized knots is called informative lattice [7].

In addition, right and left transformations are introduced, which act only on the right or, respectively, the left part of the functional number [7]. Also, such a concept as the rank of transformations plays an important role [7]. Supermodular arithmetic was created to work with these elements [7].

These transformations may have various natures. It may be all possible transformations [7].

Only 15 minimal main types of generalizing mathematical transformations exist [7]. Strictly speaking, only nine types of transformations are mathematical. The other six types of transformations are not mathematical in the classical sense of the word. Thus, the classification allows from a single point of view to classify transformations (generalized mathematical ones) that can relate to different fields of knowledge and culture.

The manifold of generalized knots is called informative lattice [16].

The theory of information computing was created as a universal theory of computation for such branches of computer science as computer arithmetic, the theory of semi-numerical algorithms, etc., [16]. So, at the time of its creation, there were not even criteria for evaluating the effectiveness of semi-numerical algorithms. This problem is one of the central ones in modern computing science.

Later, this theory was introduced in the structure of polymetrical analysis.

The main concept of this theory is information calculation: in one form or another, it is the number of mathematical operations that must be performed to solve the given problem.

Three types of informative calculation (computability) were selected:

1. Informative calculation (computability) C – the number of possible mathematical operations that are required for the resolution of the proper problem.
2. Technical informative calculation (computability)

$$C_t = C \sum_{i=1}^{n} t_i, \tag{1.2}$$

   where t_i – realization time of the proper computation.
3. Generalizing technical informative calculation (computability)

$$C_g = k_{AC} C_t, \tag{1.3}$$

   where k_{AC} – a coefficient of algorithmic complexity [1, 7].

The main principle of this theory is the *principle of optimal (minimal) informative calculations* [7]: any algebraic, including constructive, informative problem has optimal resolution for minimum informative computability C, technical informative computability C_t, or generalizing technical informative computability C_g.

This principle is analogous to the action principle in mechanics and the second law of thermodynamics. Thus, the theory of information computing in an epistemological sense is similar to analytical mechanics and thermodynamics. However, there is a significant difference. If physical theories are described mainly by continuous mathematics, then computer science is mainly described by discrete mathematics. To combine many theories into a single whole, it is advisable to use de Broglie's formulation from the thermodynamics of a point [6, 19]

$$\frac{S_a}{\mathrm{h}} = \frac{S_e}{k_B} = S_g, \tag{1.4}$$

(equivalence of quantity of ordered and disordered information [6, 7]), where S_a – action, h – Planck (Dirac) constant, S_e – entropy, and k_B – Boltzmann constant, and S_g may be represented as dimensionless measure of information. Therefore, we can go from dimensional quantities (action and entropy) to dimensionless quantities – number of proper mathematical operations. This formula allows uniting an information theory and physics in one system.

The main principles and laws of corresponding physical and information theories may be represented in the following form:

$$\delta S_g \geq 0. \tag{1.5}$$

With the equality (=) sign in eqn (1.5), we have the generalized second law of equilibrium thermodynamics and the action principle, and with the greater than (>) sign we have Shannon's theorem, the principle of non-equilibrium thermodynamics and the criterion for the existence of open systems [8]. It should be noted that in this case, this criterion can be extended to deterministic systems (the theory of Yu. Klimontovich includes the entropy functional) [8].

Thus, this interpretation of de Broglie's formula (1.4) and formula (1.5) allows us to unify and write down the main variational principles of physics (action and entropy), the condition for the existence of open systems according to Yu.A. Klimontovych and Shannon's theorem of information theory using one formula and connect them with information calculations [8]. Roughly speaking, we count elementary processes: physical and informational.

Formula (1.4) indicates a synthesis of theoretical physics and information theory. S. Wolfram draws attention to the convergence of physics and computer science [22].

A hybrid theory of systems was built for systematizing the procedure of formalization and calculations on informative lattice and may be expanded on other systems [7]. At the same time, we used Condillac's principle that the system should have a minimum of principles, preferably one [12]. We came up with two of them: the principle of combining elements into a system – the *criterion of reciprocity*, and the principle of optimality of this arrangement – the *criterion of simplicity*.

Strictly speaking, the second criterion of the hybrid theory of systems can be the principle of optimal information calculations. This makes it possible to use this theory both for the analysis of calculations in a more narrow sense, such as computed algorithms, and in a broader sense for the creation of a universal theory of computer science.

When building a universal theory of the foundations of mathematics, one should start with the problem that led to the crisis in the foundations of mathematics, which is, in a systemic sense, the problem of completeness. In other words, it is necessary to build a systematic-type theory that would simultaneously satisfy the six criteria formulated above, to one degree or another being related to the completeness and procedure of calculations in the most general sense.

For a more full formalization, the all-famous regions of knowledge and science, the *parameter of connectedness* σ_t was introduced. This parameter is characterized the number of different bonds the one element of mathematical theory with other elements of this theory. For example, in classic mathematics $\sigma_t = 1$, in linguistics, and semiotics $\sigma_t > 1$. The need to take connectivity into account when analyzing complex systems was pointed out by J. Casti [13]. Taking this fact into account, we introduced σ_t. It should be noted that connectivity to one degree or another is specified in generalized transformations as well. In general, it should be separated into a separate category of values, which is what we did.

For systematization, the computational processes, and procedures on information lattices, the hybrid theory of systems was created. This theory is based on two principles:

The *criterion of reciprocity* for the corresponding systems is the criterion for the layout (construction) of these systems. Its main features are completeness; balance; equality of the number of epistemologically equivalent known and unknown concepts [7].

Roughly speaking, the concepts of completeness and the equilibrium and the equality of the number of epistemological equivalent known and unknown notions are the nuances of the concept of completeness in methodological, concrete scientific, and mathematical senses.

The *criterion of the simplicity* for the relevant systems is the criterion of preservation in these systems of such categories [7] as completeness; balance and implementation of the principle of optimal information calculations.

As can be seen from the definitions of criteria the reciprocity and the simplicity, the first two provisions are the same. In the criterion of simplicity, only the mathematical part (principle of optimal information calculations) is clarified and strengthened.

Depending on which conditions of reciprocity and simplicity criteria are fulfilled and what is the value of the parameter the connectedness, we get only 10 minimal types of knowledge formalization systems [16]:

1. A system is calling *simple* if it fulfills all the conditions of the reciprocity and simplicity criteria for all elements of the mathematical construct (functional numbers $N_{\varphi_{ij}}$ and transformations) and the parameter of the connectedness σ_t is equal to 1.
2. The system is called *parametrically simple* if it fulfills all conditions of reciprocity and simplicity criteria only for functional numbers $N_{\varphi_{ij}}$ and the parameter of the connectedness σ_t is equal to 1.
3. The system is called *functional simple* if it fulfills all conditions of reciprocity and simplicity criteria only general mathematical transformations and the parameter of the connectedness σ_t is equal to 1.
4. System is calling *semisimple* if the principle of optimal information calculations is not preserved for it and $\sigma_t = 1$.
5. System is calling *parametric semi-simple* if the principle of optimal information calculations is not preserved only for $N_{\varphi_{ij}}$ and $\sigma_t = 1$.
6. System is calling *functional semi-simple* if the principle of optimal information calculations is not preserved only for general mathematical transformations and $\sigma_t = 1$.
7. System is calling *complicated system semisimple* if the principle of optimal information calculations is not preserved for it and $\sigma_t \neq 1$.
8. System is calling *parametric complicated* system if the principle of optimal information calculations is not preserved only for $N_{\varphi_{ij}}$ and $\sigma_t \neq 1$.
9. System is calling *functional complicated* if the principle of optimal information calculations is not preserved only for general mathematical transformations and $\sigma_t \neq 1$.
10. System is calling *absolute complicated* if none of the provisions of the criteria of reciprocity and simplicity are met and $\sigma_t \neq 1$.

Thus, we have a finite number of types the systems of the formalization the knowledge; at the same time, the number of implementations of specific systems can be arbitrarily large. So, from of the representing 10 types of systems, only 6 are mathematical in classical sense. This classification is based on the concept of completeness, which is also related to the problem of information complexity.

It is easy to see that polymetric analysis satisfies the six conditions that theories of this type must satisfy. This theory is an open-type theory. Concrete sciences are included here through generalized mathematical transformations and the parameter of connectedness. Thus, in the general case, we are not explicitly tied to concrete sciences. The main factor in this case is the very procedure of building a formalized system, which is primarily related to calculations and assessments, completeness and connectedness. It is on the problem of the completeness and connectedness of system elements that the systematic classification of simplicity–complexity of knowledge formalization systems is built.

The generalized constructive mathematical element (1.1) is an element of measure, and includes the procedure of measurements and evaluations. Moreover, quantitative mathematical transformations are responsible for primary measurements, and qualitative generalized mathematical transformations are responsible for secondary measurements [16]. Moreover, the quantum mechanical measurement procedure through right and left transformations is also generalized in the measurement procedure. The latter allow forecasting and evaluation of possible systems and results.

Practically, polymetric analysis is the most generalized formalization of N. Campbell's concept of primary and secondary measurements. The second edition of Campbell's work [5], which was reprinted in 1957 in New York, was called "Fundamentals of Science." After our formalization procedure, polymetric analysis can be represented as the foundations of mathematics [8]. It is appropriate to note that according to V. Arnold [6], "Mathematics is exact knowledge," while science is ordered knowledge.

The concept of connectivity plays an important role both in modern mathematics and information theory [13, 16]. In polymetric analysis, it is included in two ways: through qualitative generalized mathematical transformations and the parameter of connectedness. These two concepts have a different epistemological load: if in the first case this concept corresponds to a certain specific science that undergoes a corresponding transformation, then in the second case it corresponds to an interdisciplinary systemic connection. One way or another, these two concepts are more or less interconnected (depending on the type and complexity of the problem).

The idea of using a more versatile system analysis for the problem of the foundations of mathematics was laid in the organismic approach to the foundations of mathematics by A. N. Whitehead [18]. He well understood that the theory of the foundations of mathematics should include not only the question of substantiation of certain concepts and theories but also the procedure of building these theories. B. Russell's attempt to save the logical approach to the foundations of mathematics led to the construction of his inductive theory of logical types, which is based on the fact that in order to substantiate more complex systems, it is necessary to expand the operational base of new logics (introduce predicates). Although there are many mathematical disciplines in modern mathematics, the number of which increases with the development of mathematics, the volumes should be approached precisely from an operational point of view. In other words, we should use B. Russell's idea [10] of generalizing and expanding the operational base of mathematics not only for logic but also for all other existing and possible sections of it.

The intuitionistic approach to the foundations of mathematics, although it belongs to structural approaches, is more pragmatic from a computational point of view, logical or formal, because they are already based on several sections of mathematics that are laid in its foundation [7, 17]. However, even in this case, intuitionistic mathematical schemes are built in an inductive way. That is why it became necessary to create a deductive system that would take into account all the features of the construction of such theories, taking into account the most general properties of their construction. In other words, it was necessary to create a theory that would take into account the most general properties of these processes. From this point of view, polymetric analysis fits the role of such a theory [16].

It is the systematic approach that, combined with the ideas of Euclid, Descartes, Newton, Wolff, Euler, and Condillac [12], led to the creation of such a theory as polymetric analysis [16]. The number of logical Russell types can be arbitrarily large. In intuitionism, the number of types of construction of relevant systems is also arbitrarily large. However, as a deductive system, polymetric analysis has a finite number of types, and thus it can be considered a deductive extension of existing approaches in the basics of mathematics.

It should be noted that since the polymetric analysis, or rather the hybrid theory of systems, classifies formalized systems according to their complexity, it made it possible to solve the following problems, which are closely related to the foundations of mathematics.

First, it is the problem of the sweetness of information presentation and processing (the problem of the century in cybernetics of S. Beer [14]). According to J. Casti [13], this problem should be solved through calculations. In polymetric analysis, the complexity of calculations is determined by the completeness and incompleteness of knowledge formalization systems and the principle of optimal information numerosity. That is, it has two hypostases: the first at the level of calculations, and the second at the level of organization of these calculations. Thus, in this way, polymetric analysis can be considered as a solution to S. Beer's problem [14] in cybernetics based on J. Casti's idea [13].

Secondly, it is a problem of ordering and formalizing knowledge. This problem is the main one in modern science. The hybrid theory of systems is one of the options for solving this problem. It can be used both for the classification of existing systems of knowledge and science and for studying the evolution of the development of the relevant field of science in both directions, that is, both along the path of simplification and along the path of complication. In addition, this theory can be used as a universal expert system both for the revision of existing systems of knowledge and sciences, and for predicting the emergence of new sciences and their mathematization. That is, in this case, mathematics and its foundations are precisely, according to Archimedes, the point of support from which you can turn the whole world upside down [16].

In general, polymetric analysis can be presented as a universal theory, which is synthesized through measure, and in this sense, this theory can be considered as a revival of Pythagoreanism. In addition, it may be represented as a realization of E. Wigner's idea about the creation of universal science or metascience [8, 16].

It should be noted that the term theory of everything was introduced in physics. As a rule, they denote a unified theory of physical fields. According to J. Barrow [25], this idea is based on Newton's law of gravitation and its generalization of R. Boshkovich's law of universal forces $\left(\sim \frac{1}{r^2}\right)$. From this point of view, polymetric analysis can be presented as an extended formalization of the four Newtonian rules of inference in physics and their extension to all fields of knowledge, culture, and science [8].

As we can see, in order to create a universal theory of the foundations of mathematics, it is not advisable to use only well-known mathematical theories, but also to create new ones, which include, say, the philosophical concepts of N. Campbell [5, 20, 21] about primary and secondary measurements and E. Harris [4] (science is a polyphase system). Moreover, it is not necessary to distinguish between the so-called pure and applied mathematics. Thus, in computer science, according to A. Yershov, the main task is the formalization of the Canadian philosopher L. Hall's phrase "Everything that comes from the head is intelligent." It is only necessary to start with mathematics, that is, with calculations and completeness [8].

In addition, the basis of the work of the computer is the processor, which is mainly engaged in numerical and matrix calculations. Since the generalized constructive mathematical element is, roughly speaking, the element of the multifunctional matrix, it is safe to say that polymetric analysis can be considered as a generalized formalization of computer processor.

If we consider polymetric approach in terms of H. Kantor expression "The essence of mathematics lies in its freedom," this freedom is included in the variable measure and hierarchy [16].

1.4 Conclusion

1. A comparative analysis of the problem of foundations of mathematics is presented.
2. A brief analysis of the structural concepts of the foundations of mathematics showed the necessity of transitioning to natural approaches.
3. The necessity of creation and main peculiarities of the natural concept of the foundations of mathematics in its historical retrospective is researched.
4. Based on the previous point, the main requirements to the creation of the universal system of formalization of knowledge were formulated.
5. It is shown that Polymetric Analysis can be the basis of a natural concept of the foundations of mathematics.
6. The hybrid theory of systems is represented as a universal system of formalization of the knowledge.
7. The problem of simplicity–complexity in modern science was discussed.
8. It was shown that polymetric analysis and hybrid theory of systems may be represented as an operational concept in the foundation of computer science too.

References

[1] I. Newton, Mathematical Principles of Natural Philosophy to which is added Newton's system of the world. Published by Daniel Adee, New York, 1846.

[2] V. P. Dorofeev V., P. P. Trokhimchuck, Computer Sciences and Synthesis: Retrospective and Perspective. In: *Advanced in computer science*, vol. 13, ch. 1, Ed. Mukesh Singla. AkiNik Publications, New Delhi, 2022, pp. 1-32.

[3] K. Fisher, Descartes. MIFRIL, St. Petersburg, 1994. (In Russian)
[4] E. E. Harris, Hypothesis and perception. George Allen Unwin Humanities Press, London, 1970.
[5] N. R. Campbell, Physics: in elements. University Press, Cambridge, 1920.
[6] V. I. Arnold, What is mathematics? MTNMO, Moscow, 2002. (In Russian)
[7] P. P. Trokhimchuck, Foundations of Mathematics: Retrospective and Perspective. In: *Recent Research Trends in Mathematics*. Ed. R. Mehta, M. N. Sharma, vol. 3, ch.3, Integrated Publications, New Dehly, 2023, pp. 41-71.
[8] P. P. Trokhimchuck, Theories of Everything: Past, Present, Future. Lambert Academic Publishing, Saarbrukken, 2021.
[9] B. Russell A history of western philosophy. A Touchstone Book, Ney York a. o., 1972.
[10] B. Russel, Introduction to mathematical philosophy. Museum street, London, 1948
[11] Euclides, Elements. Green Lion Press, Santa Fe, 2002.
[12] E. B. De Condillac, Traite des systemes, Kessinger Publishing, New York, 2009 (In French)
[13] J. Casti, Connectivity, Complexity and Catastrophe in Large-scale Systems. John Wiley & Sons, New York, 1979.
[14] S. Beer, We and complexity of modern world. In: *Cybernetics today: problems and propositions*. Znaniye, Moscow, 1976, . 7 – 32 (In Russian)
[15] K. Gödel, On Formally Undecidable Propositions in Principia Mathematica and Related Systems I. Dower Publications, inc., New York, 1992.
[16] P. Trokhimchuck, Polymetrical Analysis. History, Concepts, Applications. Lambert Academic Publishing, Saarbrukken, 2018.
[17] Ruzha I. Foundations of mathematics. Kyiv: Vyshcha shkola, 1981. (In Russian)
[18] A. N. Whitehead, Science and the modern World. Pelican Mentor Books, New York, 1948.
[19] L. de Broglie, L. de Broglie, La thermodynamique de la particule isolée(ou la thermodynamique cachée des particules), Gauthier villars, Paris, 1964 (In French)
[20] N. R. Campbell Physics: in elements. University Press, Cambridge, 1920.

[21] N. R. Campbell, An account of the principles of measurements and calculations. Longman Green, London, 1928.
[22] S. Wolfram, Metamathematics: Foundations & Physicalization. Wolfram Media, Leipzig, 2022.
[23] S. C. Kleene, Introduction to metamathematics. North Holland, Amsterdam, 1952.
[24] W. Leibniz, Studies of Universal Calculus. In: *Gottfried Wilhelm Leibniz. Works in four volumes, Vol.* 3. Mysl', Moscow, 1984, pp. 533-537 (In Russian)
[25] J. Barrow, New theories of everything: the quest for ultimate explanation. – Oxford: University Press, 2007.

2

Versatile Applications of the Burgers Equation Across Diverse Disciplines

S. Mishra[1], G. Arora[1], J. Gupta[2], and Richa Rani[3]

[1]Department of Mathematics, School of Physical Sciences and Chemical Engineering, Lovely Professional University, India
[2]GNA University, School of Natural Sciences, Phagwara, Punjab, India
[3]Department of Mathematics, Guru Gorakhnath Ji Govt. College Hisar, Haryana, India
E-mail: mishrasam73@gmail.com; geetadma@gmail.com; jayagupta9295@gmail.com; richasardana7@gmail.com

Abstract

Recently, the scientific community has shown growing interest in analyzing convective–diffusive nonlinear partial differential equations. Burgers' equation, with a well-known exact solution, stands out as one of the renowned nonlinear equations with physical and mathematical significance suitable for analyzing various crucial areas. An overview of the historical background, highlighting not only the mathematical but also the physical importance of the Burgers equation's solution, is provided with a focus on current research approaches.

Keywords: Burgers equation (BEs), Ordinary differential equations (ODEs), Partial differential equations (PDEs), computational fluid dynamics (CFD), fluid mechanics

2.1 Introduction

For numerous decades, researchers have focused on the mathematical modeling of nonlinear convective–diffusive partial differential equations (PDEs) within the realm of computational fluid dynamics (CFD). This area of study has played a crucial role in investigating heat transfer and fluid mechanics. Computational fluid dynamics, widely employed for diverse engineering analyses, has a historical connection to numerical methods dating back to the time of Newton. Throughout this extended period, the scientific community has conceptualized distinct methods for ODEs and PDEs. Due to remarkable advancements in computational capacity, the speed of computers has undergone a much more rapid improvement compared to the associated costs. Computational fluid dynamics stands as a potent tool for addressing a diverse array of contemporary engineering challenges. Computational fluid dynamics not only offers quantitative analysis but also facilitates qualitative insights into a system. It proves invaluable in simulating interactions between solids and fluids. With the continual development of increasingly complex PDEs, the consideration of various real-life parameters affecting engineering systems becomes more comprehensive. A profound understanding of fluid flow mechanisms in pipes is imperative, given their widespread application in advanced engineering systems, industries, and everyday life. This knowledge is particularly crucial when considering the continuous flow of blood through veins and arteries in the human body. Biomedical systems, such as dialysis systems and artificial hearts, rely on the principle of fluid flow through pipes. Similarly, in applications where refrigerants circulate through pipes, like refrigeration and air conditioning, at a more expansive level, the understanding of PDEs is crucial, as they play significant roles in elucidating the intricacies of these physical processes, contributing to the analysis and design of contemporary engineering systems. The English mathematician Harry Bateman (1882–1946) [1] presented a PDE in his study, along with the associated ICs and BCs expressed by eqn (2.1)–(2.3).

$$v_t + v v_z = u v_{zz}, \quad 0 < z < A, 0 < t < \tau \tag{2.1}$$

$$v(z,0) = \psi(z), 0 < z < A \tag{2.2}$$

$$v(0,t) = \zeta_1(t), v(L,t) = \zeta_2(t), 0 < t < \tau. \tag{2.3}$$

Here, v represents velocity, z denotes spatial coordinate, t signifies time, and u stands for kinematic viscosity. Additionally, ψ, ζ_1, and ζ_2 are specified functions of variables that depend on the particular conditions relevant to the problem at hand. The Dutch physicist Johannes Martinus Burgers elucidated the mathematical modeling of turbulence using eqn (2.1). Over time, Burgers emerged as a prominent figure in the realm of fluid mechanics. In recognition of his substantial contributions, this equation is widely recognized as the Burgers equation (BE) [2].

The BEs stand out as one of the useful equations, incorporating both diffusive effects with nonlinearity and propagation effects. As a PDE in nonlinear form, BEs encapsulates various challenging difficulties encountered in engineering. The simultaneous presence of the nonlinear convective term vv_z and the diffusive term uv_{zz} adds a distinctive complexity to this equation. As the parameter u approaches zero, eqn (2.1) transforms into the inviscid BEs, serving as a model for nonlinear wave propagation.

The mathematical model extends beyond fluid mechanics, without internal heat generation encompassing the unsteady heat equation, and represents various natural phenomena. In the absence of velocity (v approaching zero), eqn (2.1) transforms into the heat equation. This formulation is particularly relevant when studying the temporal variation of temperature in solid materials, leading to corresponding variations in heat transfer rates within the material and at its boundaries. In practical engineering scenarios, understanding the temperature distribution becomes crucial for calculating thermal expansion, local heat transfer rates, and thermal stress at key locations. BEs, a foundational PDE in fluid mechanics, serves as an exemplary illustration for these reasons.

1. PDE solutions are widely recognized and established.
2. Conceptually, it can be viewed as a hyperbolic problem, incorporating artificial diffusion when dealing with small kinematic viscosity (u), and resembling a heat equation for very low values of v.
3. Its applicability extends to boundary layer calculations in the context of viscous fluid flow.
4. This equation serves as a benchmark problem for testing the performance of PDE solvers.
5. Its versatility is evident as it finds application in diverse fields such as the deposition of polydisperse suspensions and colloids, nonlinear wave propagation, longitudinal elastic waves in cosmology, aspects of turbulence, growth of molecular interfaces, gas dynamics, shock wave theory, and traffic flow in isotropic solids.

The inception of BEs traces back to 1915, when Bateman derived it in a physical context, while Fay later derived its series form in the acoustic framework. Fay's solution to BEs stands out as particularly intriguing. In 1940, Burgers introduced a special solution, underscoring its significance. The exploration of turbulence theory took a significant turn in 1948 with BEs [2–4]. In 1949, Lagerstrom et al. discovered that BEs could be transformed into the linear heat equation, a revelation later published by Cole in 1951. Independently, Cole–Hopf also unveiled the transformation of Burgers, a discovery credited to Hopf as well. This transformation, known as the Cole–Hopf transformation, converts BEs into a linear parabolic equation [5–6]. In 1972, Benton et al. contributed to the understanding of BEs by publishing 35 distinct solutions, complemented by two additional solutions for the IVPs and BVPs.

Morgan Michal's method was employed to identify appropriate groups for BEs, disregarding auxiliary conditions, as proposed by Ames in 1972 [7]. In 1989, Vorus delved into the study of the vortex sheet at the separation point, considering both steady flow states and sinusoidal conditions. The resulting BEs exhibited movement at a moderate pace. Vorus integrated the equation and implemented the Cole–Hopf transformation to mitigate the nonlinearity in the transformed upstream boundary condition [8]. Lastly, in 1997, Mazzia et al. transformed BEs into a system of ODEs, applying a transverse scheme in conjunction with boundary value methods [9]. Yan et al. discussed the numerical technique that works efficiently for solving BE [10]. In 1988, Olle provided the general solution for BE [11]. In 1999, Taigbenu discussed BEs by using the Green element method [12].

The Cole–Hopf transformation method, introduced by its namesakes, enables the derivation of the exact solution, yielding coefficients for the nonlinear term identical to those in BEs. Additionally, El Malek and El-Mansi [13] undertook an estimation of the solution for BEs obtained through Cole's method. According to the Figure 2.1, one of the applications is to understand turbulence and viscous flow.

2.2 Application of the Burgers Equation

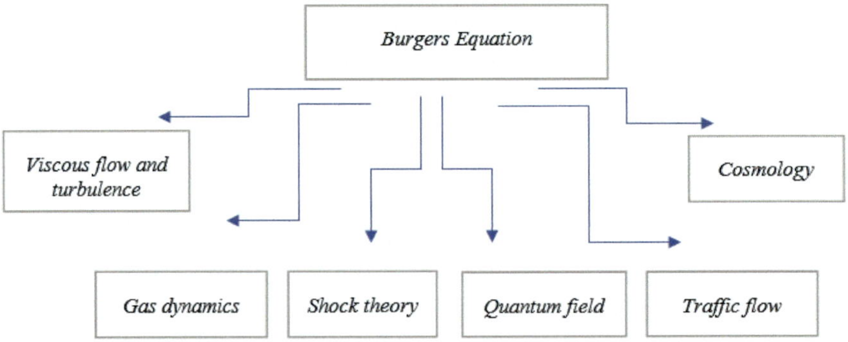

Figure 2.1 Applications of the Burgers equation.

2.2.1 Viscous flow and turbulence

To understand turbulence and viscous flow, significant strides have been made, particularly in the context of aerodynamic flow at standard pressures and temperatures. BEs, sharing key characteristics with the Navier–Stokes equation, emerge as a pivotal tool in the analysis of fluid turbulence. Murray's findings [14] revealed the existence of an ultimate, constant, turbulent state. The study described that minute disturbances eventually evolve into a singular, extensive domain.

The intersection of fluid dynamics and control theory has numerous practical applications, including the feedback control of turbulence for drag reduction. Baker et al. [15] employed a nonlinear Galerkin's method that introduces a methodology for synthesizing feedback controllers with nonlinear, finite-dimensional characteristics. Their approach utilized one-dimensional BEs with distributed control.

$$\frac{\partial Z}{\partial t} = \frac{1}{\mathrm{Re}}\frac{\partial^2 Z}{\partial x^2} - Z\frac{\partial Z}{\partial x} + b(x)v(t). \tag{2.4}$$

Consider the actuator distribution function $b(z)$. In their study, Iturriaga and Khanin [16] explored spatially periodic inviscid BEs subject to random forcing in arbitrary dimensions. They also studied the corresponding Lagrangian system, which demonstrates random-time dependence. Bogaevsky [17] demonstrated that shock discontinuities accumulate matter. Kahanin and Bec [18] interpreted Burgers' turbulence in their research as the exploration of answers to one- or multi-dimensional BEs with either random ICs or the advancement of Burgers' turbulence, which has significantly improved stochastic PDE, their invariant measures, dynamical systems theory, the study of random Lagrangian systems, and the application of field theory to comprehend dissipative anomalies and multiscale phenomena in hydrodynamic turbulence.

The findings from Chekhlov et al. [19] in a straightforward system of single dimension closely resembled the results obtained from empirical studies of real-world turbulence of three dimensions. They concentrated on examining the fluctuating velocity dynamics influenced by a time random force in BEs, uncovering a unique bi-scaling behavior [20]. Utilizing WKB calculations proved highly beneficial in comprehending the dynamics of the randomly driven BEs. Gurarie and Migdal [21] discovered the instantaneous solution, characterizing the probability distribution function associated with significant positive velocity differences in an exponential form. Yakhot and Chekhlov [22] further delved into the properties of the probability density function of velocity differences across three distinct cases. In the realm of turbulence, the departure of probability density function tails from Gaussian was construed as a manifestation of intermittency [23] of particular interest were the asymptotic characteristics of probability distribution functions linked to velocity gradients and velocity increments for solutions of the randomly forced BEs. Weinan et al. [24] utilized an innovative and direct method to investigate the asymptotic characteristics of the probability distribution of velocity gradients. The study aimed to explore the scaling characteristics of different distribution functions related to randomly forced BEs. These randomly forced BEs serve as a prototype for a deep range of challenges in statistical physics, particularly those marked by strong nonlinear effects [25].

Bec et al. [26] introduced a rapid Legendre method tailored for space-periodic Burgers' turbulence characterized by spatially smooth forcing. The researchers demonstrated that shocks possess a worldwide overarching framework dictated by the configuration space topology [27]. In a separate study, Bec and Khanin [28] introduced a variational approach to Burgers' turbulence, offering insights into the uniqueness of the primary shock and the global minimizer's existence. Gomes and its team [29] explained a straightforward and flexible method for BEs.

2.2.2 Shock theory

BEs share similarities with the Navier–Stokes equation but lacks the term that signifies the pressure factor. When kinematic viscosity is low, discontinuities in the velocity, or shocks, can occur. From a physical standpoint, the viscous model provides a more accurate explanation of real-world phenomena compared to the inviscid equation. Kreiss et al. [30] examined the viscous BEs with ICs and non-homogeneous Dirichlet BCs, aiming to investigate the impact of shocks on the convergence of solutions at steady state (S-S).

$$\begin{aligned} v_t + \frac{1}{2}v_z^2 &= \epsilon v_{zz} + f(z), t \geq 0, 0 \leq z \leq 1, \epsilon > 0 \\ v(z,0) &= g(z) \\ v(0,t) &= a, v(1,t) = b \end{aligned} \tag{2.5}$$

and the corresponding S-S problem,

$$\begin{aligned} \frac{1}{2}(y)_z^2 &= \epsilon y_{zz} + f(z), 0 \leq z \leq 1, \epsilon > 0 \\ y(0) &= a, y(1) = b. \end{aligned} \tag{2.6}$$

They explored the uniqueness, existence, and characteristics of the S-S solution, establishing that the S-S problem possesses a unique solution. For sufficiently large ϵ, it was demonstrated that the S-S equation has a solution, and, furthermore, a unique solution exists for all $\epsilon > 0$. Enhancing the convergence to the steady state is achieved by placing the shock at the boundary, while locating the shock in the interior results in considerably slower convergence. The analysis of the corresponding eigenvalue problem was employed to study the speed of convergence. Addressing problems describing the formation and propagation of shocks, the expansion of discontinuities, shock collisions, and shock-capturing schemes were utilized to solve the nonlinear BEs [31].

Inspired by the research of Kreiss et al. on the IBVP for BEs, Reyna and Ward [32] delved into the investigation of shock layer behavior associated with the viscous shock problem in the limit $\epsilon \to 0$.

$$\begin{aligned} v_t + [f(v_z)] &= \epsilon v_{zz}, t \geq 0, 0 \leq z \leq 1, \epsilon > 0 \\ v(z,0) &= v_0(z) \\ u(0,t) &= \theta, u(1,t) = -\theta. \end{aligned} \tag{2.7}$$

Here, θ is a positive constant and the convex nonlinearity $f(v)$ satisfies $f(0) = f'(0) = 0, vf'(0) > 0$ for $v \neq 0$ and $f(\theta) = f(-\theta)$ for Burgers' equation $f(v) = \frac{v^2}{2}$. Analyzing the slow shock layer motion is the main motive for finding the solution for eqn (2.7) in the limit $\epsilon \to 0$. Analytical justification for the exponentially slow phase where the shock layer drifts to its equilibrium point was given. The method of matched asymptotic expansion was utilized to describe the equilibrium solution to BEs. From symmetry, the equilibrium shock layer solution for eqn (2.7) is at $z = 1/2$. Employing this method, a one-parameter family of approximate solutions is derived.

$$v = -\theta \tanh \left[\frac{\theta \epsilon^{-1} (z - z_0(t))}{2} \right] z_0. \tag{2.8}$$

In addressing the equilibrium problem and resolving the indeterminacy in the MMAE solution for the shock layer's location, a projection method was employed. Since BEs is suited for time-dependent problems, the projection method was further applied to formulate an equation of motion governing the trajectory of the slowly advancing shock layer. The asymptotic behaviour of the solution to the equation, as the system undergoes gradual evolution.

$$\frac{dz_0}{dt} = \theta \left[e^{-\theta \epsilon^{-1} z_0} - e^{-\theta \epsilon^{-1}(1-z_0)} \right]. \tag{2.9}$$

Their primary contribution lies in demonstrating that even a slight alteration in the boundary operator can destabilize the equilibrium solution, causing the shock layer to impact one of the physical boundaries instead of smoothly reaching its static equilibrium. To analyze the behavior of the shock layer for $\epsilon << 1$, they developed a numerical method based on an initial WKB-type transformation of equation. For studying the shock layer behavior numerically with relatively small ϵ, a simple FDM can be applied to the transformed problem under well-conditioned equilibrium conditions.

2.2.3 Gas dynamics

BEs illustrate the dynamic interaction between convection and diffusion inherent in engineering problems related to viscous fluid flow [33]. Additionally, BE, incorporating source terms, has applications in aerodynamics theory [34].

$$v_t + vv_z = uv_{zz} - \lambda v \tag{2.10}$$

The significant contribution involved manipulating a problem in mathematical form derived from gas dynamics theory, specifically linked to heat exchange within the boundary layer. This is pertinent to physical applications in gas dynamics, such as in chemically reacting boundary layers. The utilization of BEs, with the inclusion of heat source terms, allows for the examination of near-equilibrium boundary layers in a gas.

$$\begin{aligned} v_t + vv_z &= uv_{zz} - \lambda v, \mathrm{z} \geq 0, \mathrm{t} \geq 0, \lambda \geq 0 \\ v(0,z) &= \varphi(z), v(t,0) = f(t). \end{aligned} \tag{2.11}$$

In eqn (2.11), the term $\lambda\ v$ denotes the heat released within the boundary layer. A numerical solution, expressed in parametric form, was derived for the BEs. The construction of solutions involved the use of Lagrange subsidiary equations and the splitting of the nonlinear PDE of order two. Consequently, the obtained solution led to the generation of Monge's equation.

2.2.4 Cosmology

In contrast to its initially uniform state, the evolving Universe exhibits irregularities and clumpiness [35]. Gurbatov and Saichev [36] explored the properties of a random potential vector field through the three-dimensional BEs. In this study, the authors delved into the potential correlation between the macroscopic Universe structure and its cellular configuration, employing the BEs in three dimensions as a theoretical framework.

$$Vt + (V_{\nabla})\, V = v\Delta V \tag{2.12}$$

$$V(m, n = 0) = V_0(m). \tag{2.13}$$

This model may find acceptance in the theory of turbulence. The behavior of the field $V(m, n)$ expressed by eqn (2.13) bears similarities to the expansion of the Universe, given that this field acquires a universal cellular structure over time, the presence and reasons for which are elucidated.

According to the author, one mechanism contributing to the formation of the Universe's cellular structure, namely inertial instability, can be examined by analyzing the statistics and dynamics of fields expressed by eqn (2.13). They also considered the potential interlink between the characteristics of these fields and the macroscopic Universe structure. The adhesion model of the large-scale distribution of matter in the Universe is formulated by Burgers' turbulence [37–38].

Following the decoupling between baryons and photons, Peebles [39] elucidated that the early Universe transitioned into a rarefied medium devoid of pressure that primarily contained non-collisional dust that was interacting through Newtonian gravity. The initial density fluctuations in this dark matter are pivotal in shaping the macroscopic structure, where concentrations of both luminous baryonic matter and dark non-baryonic matter take place. A hydrodynamical formulation provides an explanation of how matter expands with the velocity of the cosmological problem.

Molchanov et al. [40] delved into the dynamics of shock fronts in the inviscid non-homogeneous BEs in d-dimensional Euclidean space, considering the influence of random forcing due to a degenerate potential. The impact of clusters, mediated by their gravitational field, was incorporated using BEs with an external force field F.

2.2.5 Traffic flow

BEs establish a robust connection between models for traffic flow and fluid dynamics. In the theory of traffic flow, the utilization of Greenshields' model [41] gives rise to the formulation of BEs. The traffic flow on a highway is quantified as the number of vehicles per unit time passing through a reference point [42]. When the focus of vehicles is in linear form, correlated with the drift speed in the scenario of one dimension, BEs become instrumental in elucidating the vehicle concentration on the road within a moving reference framework [43].

From a refined fluid dynamical perspective, traffic jams do not spontaneously form from a uniform state of density. Leibig [44] investigated how a random initial distribution of steps in the density profile evolves over time. Treating traffic as an effectively compressible fluid in one dimension, akin to the hydrodynamic theory of fluids, facilitates the development of a "macroscopic" theory of traffic [45]. When observed from a considerable distance, traffic flow can be effectively treated as a compressible fluid (a continuum) in a single dimension [46]. Aw and Rascle [47] presented a second-order model for very light traffic flow. The density wave in traffic flow, transitioning gradually from a non-uniform to a uniform distribution, is aptly described by BEs [48]. Eqn 2.14, identified as a one-dimensional nonlinear non-homogeneous BE, finds application in diverse physical phenomena, including feedback control design [49], plasma physics [50], and wind-induced water wave buildup [51].

$$v_t + vv_z = uv_{zz} + f(z,t). \tag{2.14}$$

2.2.6 Quantum field

The BEs serve as a fundamental illustration of how a quantum computer has been notably more proficient in numerically forecasting time-dependent solutions [52]. The derivation of this equation arises from field theory in an effective manner, describing the behavior of the quantum computer on a macroscopic level, defined by infinitesimal lattice unit sizes and in step intervals of time. A quantum algorithm tailored for microscopic scales and designed for implementation on a type II quantum computer is delineated to model the time evolution of a continuous field influenced by nonlinear Boltzmann equations in a single dimension of space [53]. Within this quantum model, the system is composed of qubits, and a shorter time interval is present in the time-dependent dynamics.

In his study, Yepez [54] analyzed a quantum model in an open form that characterizes the evolution of a flow field influenced by Boltzmann equations in a single-dimensional space and is time-dependent. The effectiveness of this quantum algorithm extends to modeling highly nonlinear shock formations, even in instances of severely under-resolved shock fronts, without experiencing model breakdowns. Over time, the flow field evolving according to the Boltzmann equations (BEs) exhibits distinct, sharp features, presenting a robust testing ground for the implementation of nuclear magnetic resonance in liquid form on type II quantum computers. This stands out, particularly in comparison to examples utilizing the diffusion equation. Numerical analysis and precise analytical outputs for the Boltzmann equations affirm the practicality of applying spatial NMR techniques to implement the quantum lattice gas algorithm [55].

2.3 Conclusion

Enhancements in computational capabilities and the emergence of robust schemes are poised to invigorate multidisciplinary researchers, fostering a heightened enthusiasm for the exploration of BEs. The equation's possession of an exact solution further adds to its appeal. This chapter aims to undertake a concise discussion of historical and mathematical perspectives as well as recent advancements in the numerical simulation of BEs. In this comprehensive discussion, we earnestly underscored the significance of BEs in the context of modern engineering. We are optimistic about a promising future trajectory.

Acknowledgments

The author would like to thank the anonymous referees for their valuable feedback and the publishers for providing the opportunity to contribute to the book.

References

[1] H Bateman, "Some recent researches on the motion of fluids.", Monthly Weather Review, 1915, 43(4), pp. 163-170.

[2] J M Burgers, "Mathematical Examples Illustrating Relations Occurring in the Theory of Turbulent Fluid Motion", *Trans. R. Netherlands Acad. Sci. Amsterdam*, 1939, 17(1).

[3] JM Burgers, "A mathematical model illustrating the theory of turbulence", Advances in applied mechanics 1, 1948, pp. 171-199.

[4] FT Nieuwstadt, and JA Steketee, eds. Selected papers of JM Burgers. Springer Science & Business Media, 2012.

[5] E Hopf, "The partial differential equation [u. sub. t] + [uu. sub. x] = [[mu]. sub. xx]." Communications on Pure and Applied mathematics, 1950, 3(3), pp. 201-230.

[6] JD Cole, "On a quasi-linear parabolic equation occurring in aerodynamics." Quarterly of applied mathematics, 1951, 9(3), pp. 225-236.

[7] WF Ames, Nonlinear Partial Differential Equations, Academic Press, New York, 1972.

[8] WS Vorus, "The solution of Burgers' equation for sinusoidal excitation at the upstream boundary." *Journal of engineering mathematics*, 1989, 23, pp. 219-237.

[9] A Mazzia, Annamaria, and F Mazzia. "High-order transverse schemes for the numerical solution of PDEs." Journal of computational and applied mathematics, 1997, 82(1-2), pp. 299-311.

[10] YC Hon, and X Z Mao. "An efficient numerical scheme for Burgers' equation." Applied Mathematics and Computation, 1998, 95(1), pp.37-50.

[11] O Brander and J Hedenfalk, "A new formulation of the general solution to Burgers' equation." Wave Motion, 1998, 28(4), pp. 319-332.

[12] AE Taigbenu, "Burgers Equation", The Green Element Methods, 1999, pp. 195-216.

[13] El Malek, and El-Mansi, "Group theoretic methods applied to burger's equation", Journal of Computational and Applied Mathematics, 2000, 3-115(1-2), pp. 1-12.

[14] JD Murray, "On Burgers' model equations for turbulence", Journal of Fluid Mechanics, 1973, 59(2), pp. 263-79.

[15] J Baker, A Antonios, and CD Panagiotis, "Nonlinear control of incompressible fluid flow: Application to Burgers' equation and 2D channel flow", *Journal of Mathematical Analysis and Applications,* 2000, 252.1, pp. 230-255.

[16] R Iturriaga, and K Khanin. "Burger's turbulence and random Lagrangian systems." *Communications in Mathematical Physics,* 2003, 232, pp. 377-428.

[17] IA Bogaevsky, "Matter evolution in Burgulence", *arXiv preprint math-ph/0407073, 2004.*

[18] J Bec, and K. Khanin, "Burgers turbulence", *Physics reports*, 2007, *447*(1-2), 1-66.

[19] A Chekhlov, and Y Victor, "Kolmogorov turbulence in a random-force-driven Burgers equation." Physical Review E, 1995, 51(4).

[20] A Chekhlov, and Y Victor, "Kolmogorov turbulence in a random-force-driven Burgers equation: anomalous scaling and probability density functions." Physical Review E, 1995, 52(5).

[21] V Gurarie, and M Alexander, "Instantons in the Burgers equation", Physical Review E, 1996, 54(5).

[22] V Yakhot, and C Alexei, "Algebraic tails of probability density functions in the random-force-driven Burgers turbulence", Physical review letters, 1996, 77(15).

[23] E Balkovsky, G. Falkovich, I Kolokolov, and V Lebedev. "Intermittency of Burgers' turbulence." *Physical review letters,* 1997, 78(8).

[24] E Weinan, K Khanin, A Mazel, and Y Sinai. "Probability distribution functions for the random forced Burgers equation." *Physical review letters,* 1997, 78(10).

[25] E Weinan, E., K Khanin, A Mazel, and Y Sinai. "Invariant measures for Burgers equation with stochastic forcing." *Annals of mathematics*, 2000, pp. 877-960.

[26] J Bec, U Frisch, and K Khanin. "Kicked burgers turbulence." Journal of Fluid Mechanics, 2000, 416, pp. 239-267.

[27] J Bec, R Iturriaga, and K Khanin. "Topological shocks in Burgers turbulence." Physical review letters, 2002, 89(2).

[28] J Bec, and K Khanin. "Forced Burgers equation in an unbounded domain." *Journal of statistical physics,* 2003, 113, pp. 741-759.

[29] D Gomes, R Iturriaga, K Khanin, and P Padilla. "Viscosity limit of stationary distributions for the random forced Burgers equation." Mosc. Math. J., 2005, 5(3), pp. 613-631.

[30] G Kreiss, and H Kreiss. "Convergence to steady state of solutions of Burgers' equation." Applied Numerical Mathematics, 1986, 2(3-5), pp. 161-179.

[31] HQ Yang, and AJ Przekwas. "A comparative study of advanced shock-capturing schemes applied to Burgers' equation." *Journal of Computational Physics,* 1992, 102 (1), pp. 139-159.

[32] LG Reyna, and JW Michael, "On the exponentially slow motion of a viscous shock." Communications on Pure and Applied Mathematics, 1995, 48(2), pp. 79-120.

[33] RS Hirsh, "Higher order accurate difference solutions of fluid mechanics problems by a compact differencing technique." Journal of computational physics, 1975, 1990(1), pp. 90-109.

[34] P Peebles, *The large-scale structure of the Universe,* Princeton University Press, Princeton, 1980.

[35] SG Gurbatov, and IS Alexander, "Probability distributions and spectra of potential hydrodynamic turbulence." Radiofizika, 1984, 27(4), pp. 456-468.

[36] S N Gurbatov, A I Saichev and S F Shandarin, *Sov. Phys. Dokl.*, 1984, 38, 321.

[37] M Vergassola, B Dubrulle, U Frisch, and A Noullez. "Burgers' equation, Devil's staircases and the mass distribution for large-scale structures." Astronomy and Astrophysics, 1994, 289(2), pp. 325-356.

[38] S Albeverio, AM Stanislav, and D Surgailis. "Stratified structure of the Universe and Burgers' equation—a probabilistic approach." Probability Theory and Related Fields, 1994, 100(4), pp. 457-484.

[39] P Peebles, *Principles of physical cosmology,* Princeton University Press, Princeton, 1993.

[40] SA Molchanov, D Surgailis, and WA Woyczynski, "The large-scale structure of the Universe and quasi-Voronoi tessellation of shock fronts in forced Burgers turbulence in Rd." *The Annals of Applied Probability*, 1997, pp. 200-228.

[41] BD Greenshields, JR Bibbins, WS Channing, and HH Miller. "A study of traffic capacity." In Highway research board proceedings, 1935, 14(1), pp. 448-477.

[42] T Musha, and H Higuchi. "The 1/f fluctuation of a traffic current on an expressway." Japanese Journal of Applied Physics, 1976, 15(7).

[43] T Musha, and H Higuchi. "Traffic current fluctuation and the Burgers equation." Japanese journal of applied physics, 1978, 17(5).

[44] M Leibig, "Pattern-formation characteristics of interacting kinematic waves." Physical Review E, 1994, 49(1).

[45] R Kuhne, N Gartner, C Messner and A J Rathi (eds), *TRB Spe. Rep. 165*, 2nd edn, Traffic flow theory (1998)

[46] D Chowdhury, L Santen, and A Schadschneider. "Statistical physics of vehicular traffic and some related systems." Physics Reports, 2000, 329(4-6), pp. 199-329.

[47] A Aw, and M Rascle. "Resurrection of second order models of traffic flow." SIAM journal on applied mathematics, 2000, 60(3), pp. 916-938.

[48] X Yu, "Analysis of the stability and density waves for traffic flow." Chinese Physics, 2002, 11(11).

[49] AJ Atwell, and BB King. "Stabilized finite element methods and feedback control for Burgers' equation." In Proceedings of the 2000 American Control Conference. ACC (IEEE), 2000, 4, pp. 2745-2749.

[50] E Moreau, and O Vallée. "The Burgers equation as electrohydrodynamic model in plasma physics." arXiv preprint physics/0501019, 2005.

[51] E Moreau and O Vallee, "Connection between the Burgers equation with an elastic forcing term and a stochastic process", *Phys. Rev. E* 73, Article ID016112, 2006.

[52] J Yepez, "An efficient quantum algorithm for the one-dimensional Burgers equation." arXiv preprint quant-ph/0210092, 2002.

[53] J Yepez, "Quantum lattice-gas model for the Burgers equation." Journal of Statistical Physics, 2002, 107, pp. 203-224.

[54] J Yepez, "Open quantum system model of the one-dimensional Burgers equation with tunable shear viscosity." Physical Review A, 2006, 74(4).

[55] Z Chen, J Yepez, and GC David, "Simulation of the Burgers equation by NMR quantum-information processing." Physical Review A, 2006, 74(4).

3

Modeling Crude Oil Transport With a Variable-Diameter Pipeline System

K. Ketana and T. S. L. Radhika

Department of Mathematics, BITS Pilani Hyderabad Campus, India
E-mail: p20210113@hyderabad.bits-pilani.ac.in;
radhikatsl@hyderabad.bits-pilani.ac.in

Abstract

This study proposes a variable-diameter model for transporting crude oil through pipelines. The experimental configuration includes a pipeline system comprising three component pipes of different diameters operated at various temperatures. Leveraging available data on pipe dimensions, material (pipe) properties, crude oil characteristics, and pressure differentials at the two endpoints, we constructed a numerical model utilizing COMSOL multiphysics software. This model was subjected to simulations under varying flow and fluid conditions, generating a comprehensive dataset for subsequent analysis. To delve deeper into this dataset and discern critical patterns and influential feature sets, we employed suitable machine learning (ML) models. The application of conformalized quantile regression was instrumental in identifying the support vector regressor model as the robust ML model, enhancing the trustworthiness of the results. This model has been deployed to predict the flow rate, head loss, and power loss and compared them with those from the existing (uniform-diameter) pipeline model. The findings strongly suggest that the proposed design is ideal for oil transportation, particularly when prioritizing key parameters like power loss and head loss.

Keywords: Pipeline system, crude oil, numerical model, data analysis, machine learning models, conformalized quantile regression

3.1 Introduction

Heavy and extra-heavy oils offer a cost-effective alternative to conventional petroleum, potentially leading to more substantial refining margins and increased profits per barrel when managed effectively. Pipelines are the most convenient and economically viable means of transporting crude oils and their derivatives. Nonetheless, transporting heavy and extra-heavy crude oils presents challenges due to their limited mobility, flow characteristics, and wax and asphaltene deposition issues on pipeline surfaces. Among the technological challenges hindering the access of extra-heavy oil to energy markets, pipeline transportation stands out as a significant concern. Rafael et al. [1] identified three strategies for transporting heavy and extra-heavy oil: viscosity reduction, drag reduction, and oil upgrading, and examined existing and cutting-edge technological advancements aimed at mitigating viscosity and friction issues associated with the transportation of these crude oils from production sites to processing facilities. They emphasized the need for fundamental research to grasp basic mechanisms and the necessity to conduct pilot plant and semi-industrial experiments. Jiang et al. [2] performed numerical simulations using VOF (volume of fluid), standard k–ε, and CSS (continuum surface stress) models to replicate oil-water core-annular flow through a Π bend. Their study analyzed the impact of water fraction, flow direction, oil properties, and geometric parameters (including aspect ratio, diameter ratio, fillet radius, and roughness) on hydrodynamic performance and fouling characteristics. Their findings offered valuable insights for pipeline design, operational optimization, and Π bend structure enhancements. Reza and Nazar [3] conducted an experimental investigation to study the impact of operational variables, namely, inlet crude oil temperature, temperature difference between oil and pipe wall, crude oil flow rate, wax content, and time on wax deposition. Their study employed the Taguchi experimental design approach to assess the statistical influences of these operational variables on the wax deposition process. The study revealed that most deposition occurred within the first hour of the experiment, with only marginal increases in the response observed as the time extended, up to 3–5 hours. Gupta et al. [4] investigated axisymmetric, laminar oil-water core-annular flow within a 14.2-mm radius pipe under three conditions: vertical upward flow, vertical downward flow, and no gravitational effect. CFD (computational fluid dynamics) simulations they carried out agreed well with experimental observations. They demonstrated the capability to capture waves on the interface without imposing perturbations on the flow. Xie et al.

[5] utilized finite element simulation technology to solve a three-dimensional heat transfer model for a buried crude oil pipeline with a wax layer. Their findings indicated that the extent of thermal influence from the heated pipe is directly proportional to the oil temperature and velocity. Niazi [6] employed the core-annular flow technique for heavy oil transport and observed that this technique minimized crude oil contact with pipe walls, substantially reducing pipeline pressure drop (approximately 50%). Their study also revealed that this setup significantly reduced pumping power consumption for crude oil transportation. Ravindra Kumar et al. [7] employed CFD to investigate the influence of diameter, velocity, and temperature on heavy crude oil flow in three horizontal pipelines with inner diameters of 1, 1.5, and 2 inches and predicted flow characteristics, pressure, pressure drop, wall shear stress, shear strain rate, and friction factor. Saad et al. [8] created a rheology-based CFD model to simulate the flow of Iraqi heavy crude oil through a horizontal pipe (1-meter length, 3/4-inch inside diameter) using Ansys 15 Fluent software. Applying the power law fluid model, they observed that Iraqi heavy crude oil demonstrated non-Newtonian dilatant characteristics within a tested shear rate range of 1–40 per second. In their study, Nicolas et al. [9] formulated COPTEM (crude oil pipeline transportation emissions model), a greenhouse gas emissions model for crude oil pipeline transportation grounded in first principles and fluid mechanics. Analyzing 62 major Canadian and US pipelines exceeding 100,000 barrels/day capacity, they observed a greater variability (0.23−20.3 g CO2e/(bblÂůkm)) than previous studies. Their study identified linear velocity and pipeline diameter as crucial drivers of this variability. Francis et al. [10] developed a mathematical model incorporating water-in-oil emulsions, wax kinetics, molecular diffusion, and shear dispersion. Their model could effectively predict wax deposit growth rate, aging, and thickness in crude oil pipelines, offering valuable insights for decisions on pigging, insulation, and remediation strategies.

As we persist in refining both the design and configuration of crude oil transport pipelines, this ongoing effort advances by scrutinizing the simulated data through the lens of machine learning models. The objective is to pinpoint a reliable model that can forecast flow characteristics with high confidence. We propose to develop a pipeline system made of AISI 1020 steel with three components of different diameters and operating at diverse temperatures. The data on the steel material properties, dimensions of the component pipes, and the crude oil properties are taken from published work [11, 12]. It is important to note that the proposed geometry and the setup are multipurpose and applicable in various domains, such as optimizing shapes for ventilation

Table 3.1 Table representing the material properties of pipeline [11, 12]

Material properties	**Values**	**Units**
Density	7870	$\mathrm{Kg/m^3}$
Coefficient of thermal expansion	1.17×10^{-5}	$1/\mathrm{K}$
Poisson ratio	0.29	
Thermal conductivity	51.9	$\mathrm{W/(m.K)}$
Young's modulus	2.05×10^{11}	Pa
Bulk modulus	1.4×10^{11}	$\mathrm{N/m^2}$
Shear modulus	8×10^{10}	$\mathrm{N/m^2}$

Table 3.2 Table representing temperatures for crude oil.

Temperature (°C)	25	40	55	70	85
Viscosities (cP)	4.6	2.85	1.8	0.9	0.42

and air-conditioning ducts [13], power plants, microturbine systems, and heat exchangers [14], to mention a few. Its relevance also extends to the explosion process of methane/air-premixed gas [15].

The remaining sections of this chapter are structured as follows. Section 3.2 exposes the numerical model developed using COMSOL multiphysics software. Section 3.3 delves into the intricacies of data analysis, while Section 3.4 includes presenting results and accompanying discussions. The chapter concludes with the final section, which offers summary remarks and future directions.

3.2 Numerical Model

A three-component pipeline system was designed using the COMSOL Multiphysics software, as shown in Figure 3.1. The pipeline is constructed from AISI 1020 steel, and its material properties are detailed in Table 3.1. Each component pipe spans a length of 10 meters, with specific radii maintained at 4, 6, and 8 inches for the first, second, and third components, respectively. The junctions have been smoothed to ensure seamless transitions.

Additionally, these components are subjected to distinct temperature ranges, ranging from 25 to 55 °C. Inlet pressures are taken such that the pressure drops of 272,146, 294,299, 369,507, 392,400, 459,582, 490,500, 558,749, and 588,599 Pa (Pascal) are maintained between the ends.

The crude oil is assumed to be Newtonian with a density of 935 $\mathrm{Kg/m^3}$ and viscosity varies with temperature, as shown in Table 3.2 [11, 12].

We considered the stationary Navier–Stokes equation and implemented it using the laminar flow interface. The P1-P1(linear) finite element

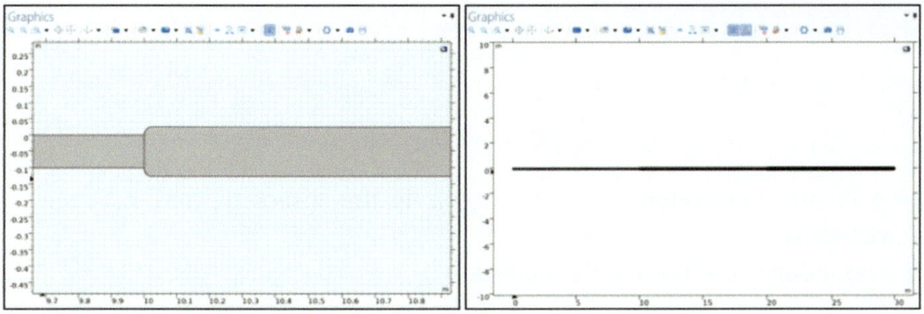

(a) (b)

Figure 3.1 COMSOL model of crude oil pipeline network with variable diameters.

Table 3.3 Parameters of the mesh.

Number of elements	2695
Number of vertex elements	14
Number of edge elements	847
Average element quality	0.8324
Minimum element quality	0.5221
Mesh area	4.571 m^2

discretization has been used for velocity and pressure. A user-controlled mesh composed of triangular (2695) elements corresponding to 5286 degrees of freedom has been used to determine the velocity and pressure variables. The mesh parameters are presented in Table 3.3.

The resulting nonlinear flow problem is solved using a nonlinear solver with no-slip conditions on the boundary walls and no backflow. The mathematical formulation of the problem is as follows:

The continuity equation is given by

$$\frac{\partial u}{\partial x} + \frac{\partial v}{\partial y} = 0. \tag{3.1}$$

Here u and v represent the fluid velocity components along x and y directions, respectively.

The equations for momentum are given by

$$\rho\left(u\frac{\partial u}{\partial x} + v\frac{\partial u}{\partial y}\right) = -\frac{\partial p}{\partial x} + \left[\frac{\partial}{\partial x}\tau_{xx} + \frac{\partial}{\partial y}\tau_{yx}\right] + \rho g_x \tag{3.2}$$

$$\rho\left(u\frac{\partial v}{\partial x} + v\frac{\partial v}{\partial y}\right) = -\frac{\partial p}{\partial x} + \left[\frac{\partial}{\partial x}\tau_{xy} + \frac{\partial}{\partial y}\tau_{yy}\right] + \rho g_y, \tag{3.3}$$

where

$$\tau = 2\mu_{\alpha pp}\mathrm{E} \tag{3.4}$$

and

$$E = \frac{1}{2}\left(\nabla u + (\nabla u)^T\right),$$

where p is the thermodynamic pressure, μ is the viscosity, and τ is the deviatoric stress

The boundary condition at the inlet is

$$u_{av}\,(x, y) = u_0 \tag{3.5}$$

where

$$u_{av} = \frac{1}{A}\int_{\partial\Omega_{\mathrm{inl}}} u.nds \tag{3.6}$$

with $A = \int_{\partial\Omega_{\mathrm{inl}}} ds.$

The outlet boundary condition is

$$p_{av}\,(x, y) = p_0, \tag{3.7}$$

where

$$p_{av} = \int_{\partial\Omega_{\mathrm{out}}} pds. \tag{3.8}$$

At the wall, a no-slip boundary condition is imposed and is given by

$$u = 0,\ v = 0. \tag{3.9}$$

3.3 Data Collection and Analysis

In this section, we outline the procedure for collecting data and describe the methodology employed for data analysis. The numerical model we developed has been subjected to simulations under various flow and fluid conditions, as outlined in Table 3.2. To explain further, we established diverse sets of conditions encompassing varying temperature-component combinations and pressure differentials and recorded the resulting data in the format illustrated in Table 3.4.

Table 3.4 Table representing the format for data collection.

S.No	T1	T2	T3	INPRE	VEL1	VEL2	VEL3	PRE1	PRE2	PRE3

3.3.1 Description of the entries

T1, T2, and T3 represent the temperatures in the first, second, and third components.

INPRE represents the inlet pressure.

VEL1, VEL2, and VEL3 represent the average velocities in the first, second, and third components.

PRE1, PRE2, and PRE3 represent the average pressures in the first, second and third component pipes.

Here, T1, T2, T3, and INPRE are the features and VEL1, VEL2, VEL3, PRE1, PRE2, and PRE3 represent the target variables.

3.3.2 Data analysis

To begin with, we performed a univariate analysis wherein we computed the descriptive statistics of the features and target variables and presented them in Table 3.5.

The next step involves multivariate analysis, generating a heat map for the feature and target variables, as depicted in Figure 3.2. This map reveals a linear relationship between the velocity (VEL1) of the first pipeline component and the pressure drop, characterized by a strong correlation index of 0.9. This signifies that as the pressure drop escalates, the velocity within the first component of the pipeline also rises. Additionally, it is important to note that the velocity demonstrates a negative correlation with the temperature of the first component pipe. This implies that as the temperature experiences a rapid increase, there is a reduction in velocity.

The variance inflation factor computed for the predictor variables revealed that the coefficient value is close to 1, as shown in Table 3.6. This suggests that the variables exhibit a low correlation with one another. Given the absence of substantial correlations among the independent variables, our initial approach employs multivariate linear regression models.

3.3.3 Linear regression model

We employed multiple linear regression on the dataset, allocating 70% as the training set, with the remaining 20% designated for testing and shown in expressions 10–12. The analysis revealed that variables T1 and PD significantly influence the velocity (VEL1) of the first component. In contrast, T2

Table 3.5 Descriptive statistics of features and exposures.

Descriptive statistics	**T 1** [°C]	**T 2** [°C]	**T 3** [°C]	**INPRE** [MPa]	**VEL1** [m/s]	**VEL2** [m/s]	**VEL3** [m/s]	**PRE1** [MPa]	**PRE2** [MPa]	**PRE3** [MPa]
Count	1000	1000	1000	1000	1000	1000	1000	1000	1000	1000
Mean	55	55	55	4.282	9.893	5.562	4.270	4.061	1.208	1.056
Std	21.22	21.22	21.22	1.088	1.866	0.942	0.723	0.820	0.063	0.012
Min	25	25	25	2.721	6.216	3.672	2.819	2.627	1.099	1.033
25%	40	40	40	3.507	8.578	4.854	3.726	3.305	1.159	1.045
50%	55	55	55	4.259	9.851	5.456	4.189	4.021	1.196	1.052
75%	70	70	70	5.075	11.306	6.258	4.804	4.673	1.253	1.064
Max	85	85	85	5.886	13.667	7.487	5.748	5.646	1.353	1.084

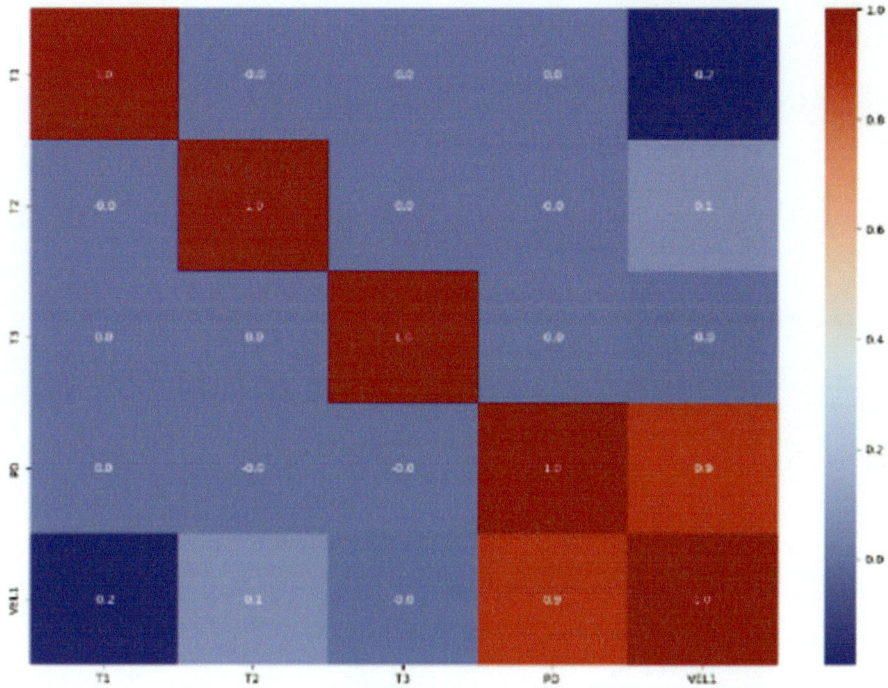

Figure 3.2 Heat map.

Table 3.6 VIF scores.

Feature	VIF score
T1	1.000
T2	1.000
T3	1.000
INPRE	1.000

and T3 exhibit less significance in this context (Table 3.7). Similar interpretations can be made on the crude oil velocity in the second and third component pipes

The linear regression model is given by

$$\text{VEL1} = -0.0009 - 0.7777\ \text{T1} - 0.0016\ \text{T2} + 0.0006\ \text{T3} + 0.5873\text{INPRE} \quad (3.10)$$

$$\text{VEL2} = -0.0009 - 0.7099\ \text{T1} - 0.0019\ \text{T2} + 0.0003\ \text{T3} + 0.6744\ \text{INPRE} \quad (3.11)$$

$$\text{VEL3} = -0.0009 - 0.7099\ \text{T1} - 0.0015\ \text{T2} - 0.000005\ \text{T3} + 0.6744\ \text{INPRE} \quad (3.12)$$

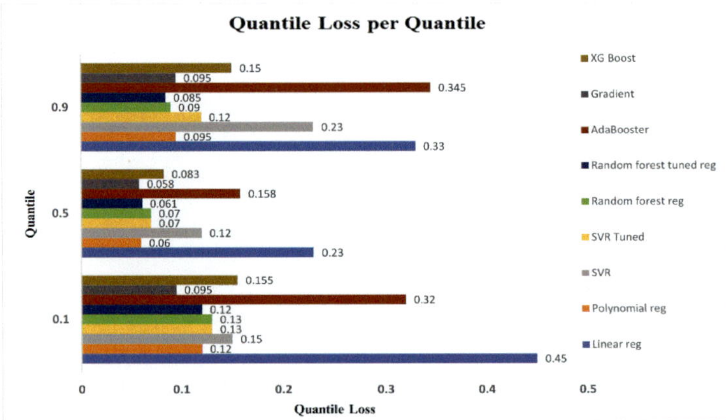

(a) Quantile loss of VEL1

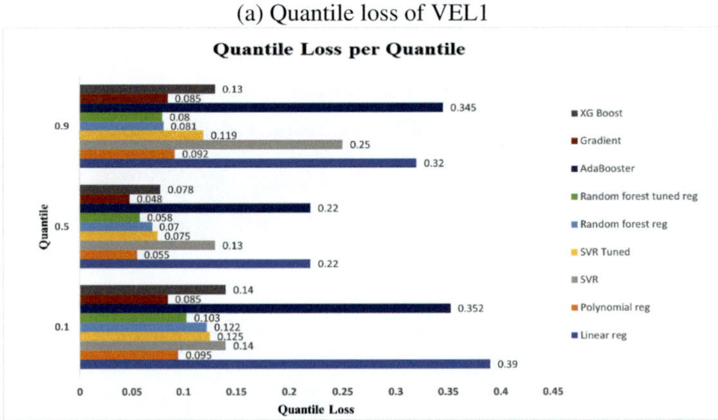

(b) Quantile loss of VEL2

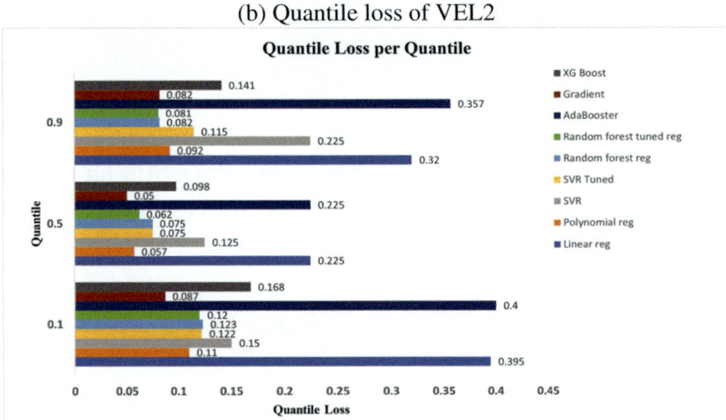

(c) Quantile loss of VEL3

(continued)

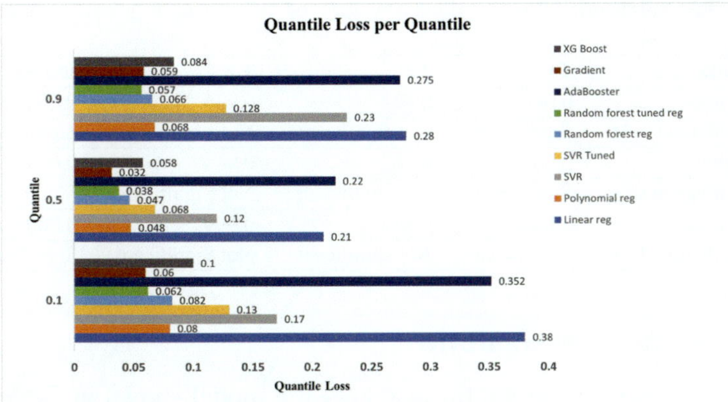

(d) Quantile loss of PRE1

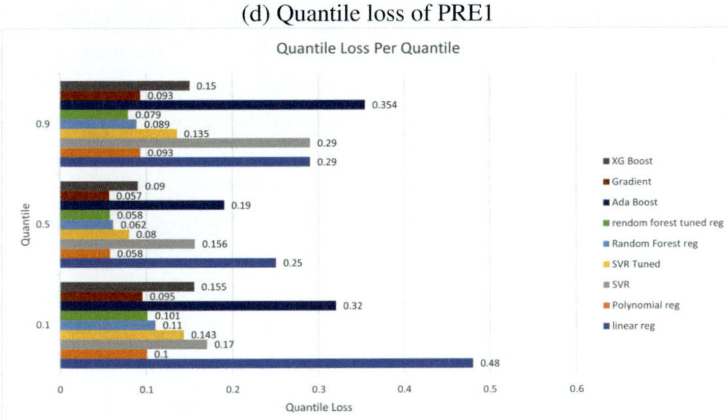

(e) Quantile loss of PRE2

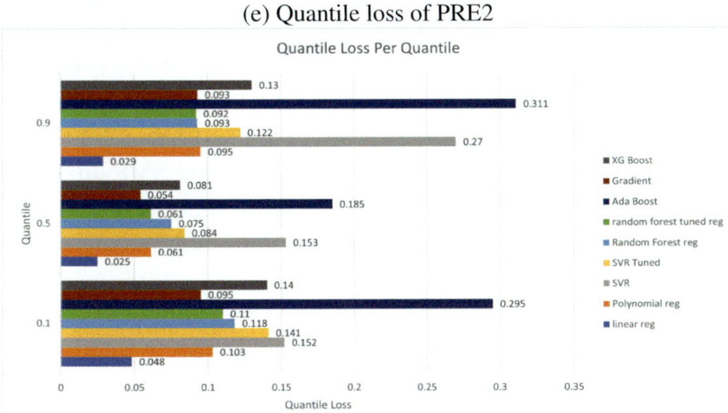

(f) Quantile loss of PRE3

Figure 3.3 Plot representing quantile loss of all ML models.

Table 3.7 Table representing the significant values of the linear regression model of velocities

VEL	Parameters	T1	T2	T3	INPRE
VEL1	*P*-values	0.000	0.822	0.936	0.000
	Significance test	Significant	Not significant	Not significant	Significant
VEL2	P-values	0.000	0.767	0.964	0.000
	Significance test	Significant	Not significant	Not Significant	Significant
VEL3	P-values	0.000	0.822	0.999	0.000
	Significance test	Significant	Not significant	Not Significant	Significant

3.4 Results and Discussions

The performance metrics on the linear regression model shown in Table 3.8 indicate that the feature set could predict the target variables accurately 95% of the time. However, we extended our data analysis to include nonlinear regression models, exploring their potential to enhance velocity predictions beyond the capabilities of linear models. Accordingly, we employed various models on the dataset, including polynomial regression, support vector regressor (SVR), decision tree regressor, AdaBoost, Gradient Boosting, and XGBoost regressor. We computed and consolidated the performance metrics for velocity in the three component pipes with an 80–20 data split, as depicted in Table 3.8. The results reveal that the XGBoost model attains the lowest mean absolute error (MAE) for the target variables VEL1, VEL2, and VEL3.

Table 3.9 further presents the performance metrics for various machine learning models developed to forecast pressure in the three components. Once again, the XGBoost model stands out with the least MAE, affirming its effectiveness in predicting the target variables PRE1, PRE2, and PRE3. However, we computed the quantile loss function for the above models to check the loss incurred in predicting the values [16, 17]. Figure 3.3 represents the quantile loss incurred by the models, revealing that the SVR-tuned model consistently demonstrates a lower quantile loss than other techniques in all target variables except PRE3, for which the linear model has arrived as the best model.

Using the data projected from this model, we conducted a performance analysis, comparing the proposed variable-diameter pipeline model against the current uniform-length model for crude oil transport. We computed the head loss and power loss using the formulae presented in eqn (3.13)–(3.16).

Table 3.8 Performance metrics for different ML models on velocity.

		R2		Adj_R2		RMSE		MAE	
Target variable	**ML model**	**Train**	**Test**	**Train**	**Test**	**Train**	**Test**	**Train**	**Test**
VEL1	**Linear**	0.957	0.963	0.957	0.962	0.205	0.194	0.166	0.155
	Fifth Degree Polynomial	0.998	0.998	0.998	0.998	0.038	0.039	0.028	0.030
	SVR	0.994	0.994	0.994	0.994	0.071	0.076	0.060	0.063
	SVR Tuned	0.996	0.996	0.996	0.996	0.061	0.063	0.053	0.053
	Random Forest	0.999	0.998	0.999	0.998	0.018	0.043	0.013	0.032
	Random Forest Tuned	0.998	0.998	0.998	0.998	0.033	0.040	0.025	0.030
	AdaBoost	0.959	0.970	0.958	0.969	0.201	0.174	0.177	0.151
	Gradient Boosting	0.998	0.998	0.998	0.998	0.039	0.039	0.028	0.029
	XGBoost	0.999	0.997	0.999	0.997	0.015	0.048	0.011	0.036
VEL2	**Linear**	0.965	0.969	0.965	0.970	0.185	0.174	0.148	0.138
	Fifth Degree Polynomial	0.998	0.998	0.998	0.998	0.035	0.035	0.026	0.027
	SVR	0.995	0.994	0.995	0.994	0.068	0.073	0.057	0.060
	SVR Tuned	0.996	0.996	0.996	0.996	0.061	0.060	0.053	0.050
	Random Forest	0.999	0.998	0.999	0.998	0.017	0.039	0.012	0.029
	Random Forest Tuned	0.999	0.998	0.999	0.998	0.031	0.036	0.023	0.027
	AdaBoost	0.955	0.966	0.955	0.966	0.210	0.184	0.184	0.158
	Gradient Boosting	0.998	0.998	0.998	0.998	0.035	0.034	0.025	0.026
	XGBoost	0.999	0.998	0.999	0.998	0.015	0.044	0.011	0.032
VEL3	**Linear**	0.965	0.970	0.965	0.969	0.185	0.174	0.148	0.138
	Fifth Degree Polynomial	0.998	0.998	0.998	0.998	0.035	0.035	0.026	0.027
	SVR	0.995	0.994	0.995	0.994	0.068	0.073	0.057	0.060
	SVR Tuned	0.996	0.996	0.996	0.996	0.061	0.060	0.053	0.050
	Random Forest	0.999	0.998	0.999	0.998	0.017	0.039	0.012	0.029
	Random Forest Tuned	0.999	0.998	0.999	0.998	0.031	0.036	0.023	0.027
	AdaBoost	0.948	0.960	0.947	0.959	0.226	0.202	0.199	0.176
	Gradient Boosting	0.998	0.998	0.998	0.998	0.035	0.034	0.025	0.026
	XGBoost	0.999	0.997	0.999	0.997	0.014	0.045	0.010	0.033

Head loss (h_f) is related to the Fanning friction factor f and is given by

$$h_f=f\left(\frac{LV^2}{2Dg}\right). \tag{3.13}$$

Here, L is the length of the pipe (m), V is the average velocity (m/s), D is the diameter of the pipe (m), and g is the gravitational force given by $9.8067\ \mathrm{m/s^2}$. For laminar flow, the friction factor is calculated using the

Table 3.9 Performance metrics for different ML models on pressure.

		R2		Adj_R2		RMSE		MAE	
Target variable	**ML model**	**Train**	**Test**	**Train**	**Test**	**Train**	**Test**	**Train**	**Test**
PRE1	**Linear**	0.964	0.968	0.964	0.967	0.187	0.181	0.157	0.147
	Fifth Degree Polynomial	0.999	0.999	0.999	0.999	0.021	0.021	0.013	0.015
	SVR	0.996	0.995	0.996	0.995	0.062	0.068	0.051	0.053
	SVR Tuned	0.996	0.996	0.996	0.996	0.054	0.057	0.047	0.050
	Random Forest	0.999	0.999	0.999	0.999	0.010	0.023	0.006	0.016
	Random Forest Tuned	0.999	0.999	0.999	0.999	0.018	0.020	0.012	0.015
	AdaBoost	0.969	0.971	0.969	0.971	0.173	0.170	0.141	0.136
	Gradient Boosting	0.999	0.999	0.999	0.999	0.021	0.019	0.013	0.013
	XGBoost	0.999	0.999	0.999	0.999	0.011	0.026	0.007	0.017
PRE2	**Linear**	0.952	0.959	0.951	0.958	0.217	0.206	0.183	0.173
	Fifth Degree Polynomial	0.998	0.998	0.998	0.998	0.033	0.035	0.024	0.026
	SVR	0.995	0.994	0.994	0.994	0.070	0.077	0.058	0.060
	SVR Tuned	0.996	0.996	0.996	0.996	0.062	0.063	0.053	0.052
	Random Forest	0.999	0.998	0.999	0.998	0.016	0.038	0.011	0.029
	Random Forest Tuned	0.999	0.998	0.999	0.998	0.030	0.036	0.022	0.027
	AdaBoost	0.959	0.966	0.959	0.965	0.199	0.187	0.174	0.162
	Gradient Boosting	0.998	0.998	0.998	0.998	0.034	0.035	0.024	0.026
	XGBoost	0.998	0.998	0.998	0.998	0.013	0.045	0.009	0.032
PRE3	**Linear**	0.951	0.959	0.951	0.958	0.217	0.208	0.184	0.174
	Fifth Degree Polynomial	0.998	0.998	0.998	0.998	0.034	0.035	0.024	0.026
	SVR	0.995	0.994	0.995	0.994	0.069	0.078	0.058	0.060
	SVR Tuned	0.996	0.996	0.996	0.996	0.061	0.063	0.052	0.052
	Random Forest	0.999	0.998	0.999	0.998	0.017	0.038	0.012	0.029
	Random Forest Tuned	0.999	0.998	0.999	0.998	0.016	0.037	0.012	0.028
	AdaBoost	0.970	0.975	0.970	0.975	0.170	0.159	0.146	0.135
	Gradient Boosting	0.998	0.998	0.998	0.998	0.035	0.036	0.025	0.027
	XGBoost	0.999	0.998	0.999	0.998	0.018	0.042	0.013	0.031

formula

$$f=\frac{16}{R_e}. \tag{3.14}$$

Here R_e is the Reynolds number and is given by

$$R_e=\frac{DV\rho}{\mu}. \tag{3.15}$$

Here, ρ is the density of the fluid and μ is the dynamic viscosity (cP). Power loss is given by

$$\text{Power loss} = \Delta P \times \frac{V\pi D^2}{4}. \tag{3.16}$$

Here ΔP represents the pressure difference (Pa) between the two ends of the pipe. The comparative results when the crude oil is maintained at 25 and 40 ^{o}C throughout the length of the pipe are outlined in Table 3.10.

Table 3.10 Performance comparison between the existing and the proposed model.

		Uniform length model [11]	Proposed model
Temp = 25 °C	Velocity [m/s]	26.328	6.228
	Pressure difference [Pa]	272,146	272,146
	Flow rate [m^3/s]	0.480	0.113
	Head loss [m]	0.138	0.032
	Power loss [W]	130,760	30,934
Temp = 40 °C		**Uniform length model**	**Proposed model**
	Velocity [m/s]	26.312	5.977
	Pressure difference [Pa]	272,146	272,146
	Flow rate [m^3/s]	0.480	0.109
	Head loss [m]	0.084	0.019
	Power loss [W]	130,675.4	29,685.82

Results indicate that, despite the lower flow rate in the proposed model, it showcased reduced head loss and power loss when contrasted with the existing model.

3.5 Conclusion

This chapter focuses on a pipeline network comprising three interconnected segments with variable diameters. The component-pipe lengths were fixed, enabling a comprehensive examination of each segment's velocity and pressure characteristics. Machine learning techniques were harnessed to forecast the behavior of these profiles under varying temperature and pressure conditions. Our findings underscore the significant influence of pressure differences and the temperature of the first component pipe in determining the velocities and pressures in each segment. In contrast, temperatures maintained in other segments exhibit minimal impact on the target variables. The nonlinear model (SVR tuned) outperformed other modeling approaches in predicting velocities, as evidenced by their lower quantile loss. Also, the linear model has lesser quantile loss for pressure in segment three.

Consequently, our study leads us to conclude that, for short-distance pipeline networks, the pressure difference and temperature or viscosity in the first component pipe assume pivotal roles in regulating the crude oil's velocity. Observations from Table 3.10 highlight a decrease in power and head loss along the pipeline when utilizing a variable diameter compared to a constant diameter model. This implies a more consistent energy loss along the pipeline with variable diameters, enhancing operational efficiency by preserving a more significant portion of energy for its intended use. The reduced head loss also contributes to cost savings and maintains steady pressure, promoting the pipeline's durability.

However, the study's limitations include assumptions of uniform component-pipe lengths, modeling crude oil as a Newtonian fluid with constant viscosity and density, and presuming laminar flow. These aspects will be addressed in our forthcoming research work. Furthermore, since the quantile loss for the proposed machine learning methods is on the order of 10^{-2}, our future effort is to delve into deep learning models to enhance the precision of predictions.

References

[1] Rafael Martínez-Palou, Maria de Lourdes Mosqueira, Beatriz Zapata-Rendon, Elizabeth Mar-Juarez, Cesar Bernal-Huicochea, Juan de la Cruz Clavel-Lopez, Jorge Aburto, 'Transportation of heavy and extra-heavy crude oil by pipeline: A review', Journal of Petroleum Science and Engineering, 75(3–4), 2011, 274-282.

[2] Fan Jiang, Yijun Wang, Jiajie Ou, Zhongmin Xiao, 'Numerical Simulation on Oil–Water Annular Flow through the Π Bend', Industrial & Engineering Chemistry Research, 53 (19), 2014, 8235-8244.

[3] Reza Valinejad, Ali R. Solaimany Nazar, 'An experimental design approach for investigating the effects of operating factors on the wax deposition in pipelines', Fuel, 106, 2013, 843-850.

[4] R. Gupta, C. K. Turangan, R. Manica, 'Oil-water core-annular flow in vertical pipes: A CFD study', Can. J. Chem. Eng., 94, 2016, 980-987.

[5] H Xie, C Li, W Jia, C. Zhang, 'Three-dimensional CFD Modeling on the Thermal Characteristics of Buried Oil Pipeline Involving the Heat Transfer of Wax Layer', Energies, 15(16), 2022, 6022.

[6] S Niazi, 'CFD Simulation of a Crude Oil Transport Pipeline: Effect of Water', Petroleum & Petrochemical Engineering Journal, 2, Aug., 2018.

[7] Ravindra Kumar, Shirsendu Banerjee, Anirban Banik, Tarun Kanti Bandyopadhyay, Tarun Kumar Naiya, 'Simulation of single phase non-Newtonian flow characteristics of heavy crude oil through horizontal pipelines', Petroleum Science and Technology, 35(6), 2017, 615-624.

[8] Saad Nahi Saleh, Thamer Jasim Mohammed, Huda K. Hassan, Shahzad Barghi, 'CFD investigation on characteristics of heavy crude oil flow

through a horizontal pipe', Egyptian Journal of Petroleum, 30 (3), 2021, 13-19.

[9] Nicolas Choquette-Levy, Margaret Zhong, Heather MacLean, and Joule Bergerson, 'COPTEM: A Model to Investigate the Factors Driving Crude Oil Pipeline Transportation Emissions', Environmental Science & Technology, 52 (1), 2018, 337-345.

[10] Francis Oketch Ochieng, Mathew Ngugi Kinyanjui, Jeconia Okelo Abonyo, Phineas Roy Kiogora, 'Mathematical Modeling of Wax Deposition in Field-Scale Crude Oil Pipeline Systems', Journal of Applied Mathematics, 2022, 2022, Article ID 2845221, 13 pages.

[11] R. S. Hamied, A. N. Mohammed Ali, K. A. Sukkar, 'Enhancing Heavy Crude Oil Flow in Pipelines through Heating-Induced Viscosity Reduction in the Petroleum Industry', Fluid Dynamics & Materials Processing, 19(18), 2023, 2027–2039.

[12] https://www.matweb.com/Search/MaterialGroupSearch.aspx?GroupID =229

[13] Ran Gao, Meng Wang, Wei Guo, Sikai Zhang, Shihao Wen, Dingmeng Wu, Hang Zhou, Angui Li, 'Study of the shape optimization of a variable diameter in a ventilation and air-conditioning duct based on the Boltzmann function', Journal of Building Engineering, 43, 102833, 2021.

[14] H Gül, D Evin, 'Experimental Investigation of the Effect of the Cross-Sectional Geometry on the Flow', Gazi University Journal of Science, March., 2010, 20(1), 7-14.

[15] Jian Wang, Ziyao Fan, Yong Wu, Ligang Zheng, Rongkun Pan, and Yan Wang, 'Effect of Abrupt Changes in the Cross-Sectional Area of a Pipe on Flame Propagation Characteristics of CH4/Air Mixtures', ACS Omega, 6(23), 2021, 15126-35.

[16] Xi Chen, et. al., 'Distributed High-dimensional Regression Under a Quantile Loss Function', Journal of Machine Learning Research, 21, 2020, 1-43.

[17] T. Raja Rani, Woshan Srimal, Abdullah Al Shibli, Nooh Zayid Suwaid Al Bakri, Mohamed Siraj, T. S. L. Radhika, 'Quantile Loss Function Empowered Machine Learning Models for Predicting Carotid Arterial Blood Flow Characteristics', WSEAS Transactions on Biology and Biomedicine, 20, 2023.

4

An Alternate Approach to Solve Two-Stage Transportation Problems With Uncertain Parameters

Jajji Singla[1], Gourav Gupta[2], Manoj Kumar[3], Sukhveer Singh[3], and Deepika Rani[4]

[1]Chitkara University Institute of Engineering and Technology, Chitkara University, India
[2]Department of Mathematics, Lovely Professional University, India
[3]Department of Mathematics, Graphic Era Hill University, India
[4]Department of Mathematics, Dr. B. R. Ambedkar National Institute of Technology, India
E-mail: jajjisingla.js@gmail.com; gouravgupta333@gmail.com; docmanojkumar64@gmail.com; sukhveer.singh@thapar.edu; ranid@nitj.ac.in

Abstract

The transportation problem is among the most significant categories of linear programming problems. This coordinates the movement of an item from its place of origin to its final destination. The primary objective is to meet destination requirements while minimizing freight expenses. The two-stage uncertain transportation cost problem presented in this chapter and parametric approach is used to obtain a fuzzy solution. A novel approach is suggested to address a two-phase uncertain transportation cost problem where the transportation cost is expressed in terms of uncertain trapezoidal numbers. This approach is particularly effective because it is very simple and easy to understand. The suggested method assists decision-makers with logistics-related problems by supporting their decision-making process and offering a simple, effective solution.

Keywords: Optimal solution, two-stage fuzzy transportation problem, trapezoidal fuzzy number

4.1 Introduction

The subject of optimization is a relatively recent development in mathematics. It has improved over the previous few decades due to its ability to solve issues in the real world. The goal of optimization is to find the best answer by comparing and identifying every possible solution. Initially, the fundamental transportation problem was developed by Hitchcock [5], when he presented his findings regarding the distribution of goods to various locations from multiple sources. This study is the first significant contribution to the solution of transportation-related problems. Koopman [27] conducted an independent study on the optimal use of transportation systems. The purpose of this study is to apply the theory of the optimal allocation of resources. The development of transportation models was boosted by these two contributions.

Tuncay Can's approximation method was introduced by Can and Kocak [26] as an alternative to the approximation strategies known in the literature for a balanced transportation problem. An efficient technique for determining a superior initial basic feasible solution for a balanced transportation issue was presented by Hossain et al. [15]. Hosseini [4] proposed an efficient algorithm in three cases with its Matlab code, which is forcefully efficient for problems of large sizes. Hussein and Shiker [9] introduced a new revision to the VAM and found an initial basic feasible solution for transportation problems that was almost similar to the best solution. A new and effective method, as well as an initial basic feasible solution method, was invented by numerous authors [13, 16, 19, 21, 23] for solving balanced and unbalanced transportation problems to reduce the cost of transferring a specific quantity of items from sources to destinations. In their study, Sathyavathy and Shalini [17] examined four distinct recommended means (PAM, PHM, PGM, and PQM) and concluded that the primary requirement for transportation problems is the best solution to a transportation problem. Singla et al. introduced a revision of the Vogel approximation method algorithm [12] that consistently yields a feasible solution that is similar to the best solution.

Choudhary [3] suggested an ideal revised distribution method, which is based on distributing units across the cells in the transportation matrix, in order to find the optimal solution. This method is easy to apply to both balanced and unbalanced transportation cost problems, which maximize or minimize the objective function. Gupta et al. [7] proposed an approach called the zero suffix method that gives the optimal solution for transportation problems without initially finding the initial basic feasible solution. To deal with three-level time minimization transportation cost issues, authors Singh and Singh [8] proposed a modified version of particle swarm optimization (PSO). In this case, the path from origin to terminal points is divided into three levels according to the requirements of transport enterprises, and the goal is to objectively minimize total time. It was found that the suggested method performed better when it came to making logistical decisions than existing approaches.

The most essential mathematical tools for modeling and maintaining uncertain systems in business, the human body, and the natural sciences are fuzzy sets and fuzzy logic. Due to parameter variations, real-world problems are naturally imprecise. The characteristic function can be generalized in a way that indicates the membership level of the set for each element in the universal set and allows its values to lie within a specified range. The first uncertain system was proposed by Zadeh [14] in 1965 during his seminar work on fuzzy sets. The fuzzy set theory differs from classical set theory in that it states that an object either belongs to a set or it does not. Fuzzy sets are those that permit different membership levels in the interval [0,1]. The decision-maker in the traditional transportation model is considered to have certainty about the precise data for product demand, product capacity, and transportation costs. However, in practical applications, it may not be possible to accurately obtain all parameters due to uncontrollable factors. This is why fuzzy logic systems were created for traffic engineering. Fuzzy transportation problems are defined as those in which the cost, supply, and demand variables have uncertain numbers for any one, two, or all three of the parameters. So many researchers have presented their approaches to solving fuzzy situations in the literature. Noora and Karami [1] developed DEA models using imprecise data represented by uncertain sets. By using a ranking function, the author introduces an approach to solving the fuzzy non-racial model. In their discussion of the fuzzy transportation cost issues, Kirupavathi and Arthi [24] approximated the solution for the treatment of blood cancer using uncertain trapezoidal numbers.

Usually, real-world problems are depicted with multiple goals, and these goals have different measurement scales and conflict at the same time. In some cases, due to storage limitations, the destination cannot receive more than its minimum requirement. After consuming part of this first shipment, they are ready for the second stage, which involves receiving the surplus quantity. As a result, there are two stages involved in moving the items from the origin to the destination. In the first stage of the problem, the minimal demand of a product at a destination is transferred to the corresponding destination. They are prepared to receive the remaining quantity in the second stage after using up a part of the first shipment. This class of transportation problem is known as a two-stage transportation cost issue. Transporting items from the origin to the destination in two stages with the goal of minimizing the combined transportation costs of the two stages is the major objective of the two-stage transportation cost issues. Product transportation from the source to the destination occurs in parallel during these two stages. The zero-point method, which Pandian and Natarajan [20] introduced, is a novel approach that can solve the transportation cost issues optimally without requiring optimality test procedures or even initial feasible solutions. For the two-stage time minimization problem, Sonia and Rita Malhotra [25] established a polynomial bound approach to determine the best scheduling of the first and second stages to minimize the sum of the related transportation time. A two-stage fixed-charge transportation issue with a heuristic evolutionary algorithm that utilizes properties to lead the algorithm toward better solutions was developed by Calvete et al. [10]. By converting goods in stock and consumer demand into trapezoidal fuzzy numbers using parametric approaches, Ghani and Razak [2] reduced the two-stage uncertain transportation cost. Ritha and Vinotha [30] proposed the use of geometric planning as a method to find the best compromise solution for the more objective two-stage uncertain transportation problem. Sudhakar and Navaneetha [28] proposed an effective zero suffix strategy to solve a two-stage fuzzy transportation issue that is multi-objective. Hashmi et al. [18] proposed a multi-objective strategy for a two-stage fixed-charge transportation cost issue. Singh et al. [22] suggested a two-step method for forecasting soil nutrients. Gupta et al. [6] provided an alternate method for figuring out the first basic feasible solution for uncertain transportation prices. Determining the grade value with zero expenses yields the first answer. Singla et al. [11] suggested the initial basic feasible solution that is either similar or even closest to the best solution for transportation issues in an uncertain environment, with less computational labor. Jaiswal et al. [29] computed a heptagonal number using a range ranking function that comprised the current zero point, zero suffix, and expanded zero suffix. They found that the current zero-point technique yields a more optimal result.

4.2 Preliminaries

4.2.1 Fuzzy (uncertain) number

A fuzzy number $\mathcal{P}$ on the real line *R* needs to fulfil certain requirements as:

(i) $\mu_P(x)$ must be piecewise continuous.
(ii) At least one a∈R exists where $\mu_P(x) = 1$.
(iii) $\mathcal{P}$ should be normal and convex.

4.2.2 Trapezoidal uncertain number

If the memberships function of a fuzzy set $\mathcal{P} = (p, q, r, s)$, defined on real numbers *R*, possesses the following properties, it is considered a trapezoidal uncertain figure:

(i) $\mu_P(x)$: $X \rightarrow \{0,1\}$ is continuous.
(ii) $\mu_P(x) = 0\ \forall\ x \in (-\infty, r] \cup [s, \infty)$.
(iii) It is strictly rising on $[p, q]$ and strictly falling on $[r, s]$.
(iv) $\mu_P(x) = 1 \forall\ x \in [p, q]$.

4.2.3 Membership function of trapezoidal fuzzy number

Trapezoidal uncertain numbers like $\mathcal{P} = (p,q,r,s)$ have a membership function $\mu_P(x)$ that can be represented as follows:

$$\mu_P(x) = \begin{cases} \frac{x-p}{q-p}; & p \leq x \leq q, \\ 1; & q \leq x \leq r \\ \frac{x-r}{s-r}; & r \leq x \leq s \\ 0 & s \geq 0 \end{cases}.$$

4.2.4 De-fuzzification of trapezoidal uncertain numbers

The ranking mapping is a helpful tool for comparing uncertain numbers [1]. Every uncertain number is mapped into a real number by a ranking mapping $\Re : F(R) \longrightarrow R$, where *R* is the set of real numbers and *F*(*R*) is the set of all uncertain numbers defined on a set of real numbers. Let $\mathcal{P} = (p_1, q_1, r_1, s_1)$ and $\mathcal{Q} = (p_2, q_2, r_2, s_2)$ be two uncertain numbers, then

(i) If Rank of $(\mathcal{P}) \geq$ Rank of $(\mathcal{Q})$ then $\mathcal{P} \geq \mathcal{Q}$
(ii) If Rank of $(\mathcal{P}) >$ Rank of $(\mathcal{Q})$ then$\mathcal{P} > \mathcal{Q}$
(iii) If Rank of $(\mathcal{P}) =$ Rank of $(\mathcal{Q})$ then $\mathcal{P} = \mathcal{Q}$

Let $\mathcal{P} = (p, q, r, \text{and } s)$ be the trapezoidal uncertain number, then the rank of $\mathcal{P}$ is

Rank of $(\mathcal{P}) = \frac{p+q+r+s}{4}$.

4.2.5 Algebraic operations on trapezoidal uncertain numbers

Here arithmetic operations addition, subtraction, etc., between two trapezoidal uncertain numbers, defined on a set of real numbers *R*, are given in this section [14].

Let $\mathcal{P} = (p_1, q_1, r_1, s_1)$ and $\mathcal{Q} = (p_2, q_2, r_2, s_2)$ are trapezoidal uncertain numbers, then

(i) $\mathcal{P} + \mathcal{Q} = (p_1 + p_2, q_1 + q_2, r_1 + r_2, s_1 + s_2)$
(ii) $\mathcal{P} - \mathcal{Q} = (p_1 - s_2, q_1 - r_2, r_1 - q_2, s_1 - p_2)$
(iii) $\alpha * \mathcal{P} = \{\alpha p_1, \alpha q_1, \alpha r_1, \alpha s_1 \quad \alpha > 0$

$\alpha s_1, \alpha r_1, \alpha q_1, \alpha p_1 \ \alpha < 0$

4.3 Two-Stage Fuzzy Transportation Problem

The following is the tabular representation (Table 4.1) and linear programming formulation for the two-stage fuzzy transportation problem with uncertain cost of transportation.

Table 4.1 Tabular representation of the two-stage fuzzy transportation problem.

	W_1	W_2	...	W_n	**Supply** (s_i)
F_1	$\tilde{c}_{11}$	$\tilde{c}_{12}$	...	$\tilde{c}_{1n}$	s_1
F_2	$\tilde{c}_{21}$	$\tilde{c}_{22}$	...	$\tilde{c}_{2n}$	s_2
$\vdots$	$\vdots$	$\vdots$	$\vdots$	$\vdots$	$\vdots$
F_m	$\tilde{c}_{m1}$	$\tilde{c}_{m2}$	...	$\tilde{c}_{mn}$	s_m
Demand (d_j)	d_1	d_2	...	d_n	

Linear programming formulation of the two-stage fuzzy transportation problem is

Minimize $Z = Z_1 + Z_2$

$$\sum_{i=1}^{m}\sum_{j=1}^{n} \tilde{c}_{ij}x'_{ij} \leq Z1$$

$$\sum_{i=1}^{m}\sum_{j=1}^{n}\tilde{c}_{ij}x''_{ij} \leq Z2$$

$$\sum_{j=1}^{n} x'_{ij} \leq s_i; \quad \forall i = 1,\ 2, \ldots, m \ \ldots\ldots\ldots\ldots \ \text{(i)}$$

$$\sum_{i=1}^{m} x'_{ij} = k_j\ ; \quad \forall j = 1,\ 2, \ldots, n \ \ldots\ldots\ldots \ldots \ \text{(ii)}$$

$$\sum_{j=1}^{n} x''_{ij} = s_i - \sum_{j=1}^{n} x'_{ij}; \quad \forall\ i = 1, 2, \ldots, m \ \ldots\ldots\ldots\ldots \ \text{(iii)}$$

$$\sum_{i=1}^{m} x''_{ij} = d_j - k_j; \quad \forall\ j = 1, 2, \ldots, n \ \ldots\ldots\ldots \ldots \ \text{(iv)}$$

$$x'_{ij}, x''_{ij} \geq 0, \forall\ i \text{ and } j$$

where

m = number of sources, *n* = number of destinations, *i* = index of the sources, *j* = index of the destinations, x'_{ij} = amount of goods to be shipped from source *i* to destination *j* in stage I, x''_{ij} = quantity of goods to be shipped from source *i* to destination *j* in stage II, k_j = maximum storage capacity of destination *j*, d_j = quantity of goods required at destination *j*, and c_{ij} = fuzzy transportation cost of one unit of goods from source *i* to destination *j*.

4.4 Proposed Approach

Step 1: Consider the two-stage fuzzy transportation problems whose transportation cost is taken as fuzzy number.

Step 2(a): Check $s_i<0$ and $d_j<0$, then stop.

(b) If $\sum s_i > \sum d_j$ or $\sum s_i < \sum d_j$, balance the transportation problem by introducing the dummy column or row. Take the average of the availability and demand in stage I and remaining in stage II in such a way that the obtained transportation problem should be balanced.

Step 3(a): Determine the penalties by subtracting the minimum and next to minimum fuzzy cost, by using ranking function and algebraic operations, for each row and each column, i.e.,

$p_i = |c_{ih} - c_{ik}|$ and $p_j = |c_{ri} - c_{sj}|$, where p_i and p_j are the penalty of the *i*th row and *j*th column and $c_{ij} = \Re(\tilde{c}_{ij})$.

Step 3(b): The penalty of the associated row or column is zero if more than one cell in that row or column has the minimum cost.

Step 4: Select a row and column, say *r*th row and *c*th column, such that $p_r = \max\{p_i\}$ and $p_c = \max\{p_j\}$ and allocate the maximum possible quantity $x_{rc} = \min\{a_r, b_c\}$ to this cell. Choose the row or column with the lowest cost cell (*u*, *v*) if several rows or columns have the highest penalty.

Step 4.1: If more than one minimum cost cell exists in *r*th row or *c*th column, then select the cell (*l*, *t*) in which we can allocate maximum quantity, i.e., $x_{lt} = \max\{x_{uv}\}$.

Step 4.1.1: If more than one cell will receive the maximum allocation, choose the cell (*q*, *s*) with the greatest remainder of supply and demand, i.e., $\max\{|a_q - x_{qs}|, |b_s - x_{qs}|\} >$

$$\max\{|a_l - x_{lt}|, |b_t - x_{lt}|\}.$$

Step 5: Subtract the allotted quantity to update the availability at the sources and the demand at the destinations. If a column or row is fully satisfied, remove it. In a situation where both the row and the column are satisfied simultaneously, remove one of them, and assign zero (write ε) supply or zero (write ε) demand to the remaining one.

Step 6: Repeat the steps from Step 3 to Step 5 till all the rows and columns are satisfied.

Step 7: Repeat Steps 3–5 for stage II by taking the remaining availability and demand.

Step 8: The total minimum fuzzy cost is $\tilde{Z}_1 + \tilde{Z}_2$, where $\tilde{Z}_1$ minimum fuzzy cost is obtained at stage I and $\tilde{Z}_2$ minimum fuzzy cost is obtained at stage II.

4.5 Illustrative Examples

Example 1: Consider the following uncertain transportation problem (Table 4.2) in which the cost of transporting a single unit of goods is taken as trapezoidal uncertain number.

Table 4.2 Uncertain transportation problem.

	W_1	W_2	W_3	W_4	W_5	W_6	Supply
F_1	(2,3,2,1)	(4,1,5,2)	(4,6,7,3)	(10,9,12,13)	(6,3,5,2)	(4,0,1,3)	6
F_2	(4,6,1,5)	(8,7,9,4)	(9,8,11,8)	(7,4,5,4)	(12,9,8,11)	(7,5,0,4)	8
F_3	(11,14,13,10)	(35,30,15,20)	(12,9,8,7)	(8,5,7,4)	(20,35,25,24)	(12,15,10,11)	7
F_4	(6,7,9,10)	(7,6,8,7)	(10,12,8,6)	(20,25,30,21)	(12,11,7,10)	(9,10,6,7)	7
Demand	3	5	5	6	4	5	

For stage I (Table 4.3):

We take $s_1 = \frac{6}{2} = 3$, $s_2 = \frac{8}{2}\, 4$, $s_3 = \frac{7}{2} = 3.5 = 3$, $s_4 = \frac{7}{2} = 3.5 = 3$
$d_1 = \frac{3}{2} = 1.5 = 1, d_2 = \frac{5}{2} = 2.5 = 2, d_3 = \frac{5}{2} = 2.5 = 3, d_4 = \frac{6}{2} = 3, d_5 = \frac{4}{2} = 2, d_6 = \frac{5}{2} = 2.5 = 2.$

Table 4.3 Stage I of uncertain transportation problem.

	W_1	W_2	W_3	W_4	W_5	W_6	Supply
F_1	(2,3,2,1)	(4,1,5,2)	(4,6,7,3)	(10,9,12,13)	(6,3,5,2)	(4,0,1,3)	3
F_2	(4,6,1,5)	(8,7,9,4)	(9,8,11,8)	(7,4,5,4)	(12,9,8,11)	(7,5,0,4)	4
F_3	(11,14,13,10)	(35,30,15,20)	(12,9,8,7)	(8,5,7,4)	(20,35,25,24)	(12,15,10,11)	3
F_4	(6,7,9,10)	(7,6,8,7)	(10,12,8,6)	(20,25,30,21)	(12,11,7,10)	(9,10,6,7)	3
Demand	1	2	3	3	2	2	13

By using Steps 3–5 of the proposed approach, we get Table 4.4 as given below:

Table 4.4 After first allocation.

	W_1	W_2	W_3	W_4	W_5	W_6	Supply	p_i
F_1	(2,3,2,1)	(4,1,5,2)	(4,6,7,3)	(10,9,12,13)	(6,3,5,2) 2	(4,0,1,3)	3 1	0
F_2	(4,6,1,5)	(8,7,9,4)	(9,8,11,8)	(7,4,5,4)	(12,9,8,11)	(7,5,0,4)	4	0
F_3	(11,14,13,10)	(35,30,15,20)	(12,9,8,7)	(8,5,7,4)	(20,35,25,24)	(12,15,10,11)	3	3
F_4	(6,7,9,10)	(7,6,8,7)	(10,12,8,6)	(20,25,30,21)	(12,11,7,10)	(9,10,6,7)	3	1
Demand	1	2	3	3	2	2		
p_j	2	4	4	1	6	2		

As a result, we could receive all of the stage I allocations indicated in Table 4.5.

Table 4.5 After all allocations in stage I.

	W_1	W_2	W_3	W_4	W_5	W_6	Supply
F_1	(2,3,2,1)	(4,1,5,2)1	(4,6,7,3)	(10,9,12,13)	(6,3,5,2)2	(4,0,1,3)	1
F_2	(4,6,1,5)1	(8,7,9,4)	(9,8,11,8)1	(7,4,5,4)	(12,9,8,11)	(7,5,0,4)2	4
F_3	(11,14,13,10)	(35,30,15,20)	(12,9,8,7)	(8,5,7,4)3	(20,35,25,24)	(12,15,10,11)	3
F_4	(6,7,9,10)	(7,6,8,7)1	(10,12,8,6)2	(20,25,30,21)	(12,11,7,10)	(9,10,6,7)	3
Demand	1	2	3	3	2	2	15

The obtained solution is $x'_{12} = 1$, $x'_{15} = 2$, $x'_{21} = 1$, $x'_{23} = 1$, $x'_{26} = 2$, $x'_{33} = 0, x'_{34} = 3$, $x'_{42} = 1$, and $x'_{43} = 2$. The objective value is $\widetilde{Z}_1 = (94,\ 76, 72, 58)$.

For stage II:

We take $s_1 = 3$, $s_2 = 4$, $s_3 = 4$, $s_4 = 4$, $d_1 = 2$, $d_2 = 3$, $d_3 = 2$, $d_4 = 3$, $d_5 = 2$, $d_6 = 3$.

After applying Steps 3–5 of the proposed approach, the solution is obtained as shown in Table 4.6.

Table 4.6 After all allocations in stage II.

	W_1	W_2	W_3	W_4	W_5	W_6	Supply
F_1	(2,3,2,1)	(4,1,5,2)1	(4,6,7,3)	(10,9,12,	(6,3,5,2)2	(4,0,1,3)	3
F_2	(4,6,1,5)1	(8,7,9,4)	(9,8,11,8)	(7,4,5,4)	(12,9,8,11)	(7,5,0,4)3	4
F_3	(11,14,13,10)	(35,30,15,20)	(12,9,8,7)1	(8,5,7,4)3	(20,35,25,24)	(12,15,10,11)	4
F_4	(6,7,9,10)1	(7,6,8,7)2	(10,12,8,6)1	(20,25,30,21)	(12,11,7,10)	(9,10,6,7)	4
Demand	2	3	2	3	2	3	

The obtained solution is $x''_{12} = 1$, $x''_{15} = 2$, $x''_{21} = 2$, $x''_{26} = 3$, $x''_{33} = 1$, $x''_{34} = 3$, $x''_{41} = 1$, $x''_{42} = 2$, and $x''_{43} = 1$. The objective value is

$$Z_1 = (107,\ 83,\ 78,\ 72).$$

Therefore, the optimal cost of the two-stage fuzzy transportation problem is $\tilde{Z} = \tilde{Z}_1 + \tilde{Z}_2$.

That is, $\tilde{Z} = (94, 76, 72, 58) \oplus (107, 83, 78, 72) = (201, 159, 150, 130)$.

Example 2: Consider another balanced uncertain transportation problem as shown in Table 4.7.

Table 4.7 Balanced uncertain transportation problem.

	W_1	W_2	W_3	W_4	W_5	Supply
F_1	(2,3,4,7)	(3,2,1,2)	(3,4,2,3)	(1,4,2,1)	(5,8,4,7)	6
F_2	(7,4,6,3)	(3,4,6,3)	(8,6,4,2)	(3,2,1,2)	(2,0,1,1)	12
F_3	(8,3,5,8)	(7,6,5,2)	(5,3,1,7)	(8,7,6,7)	(5,9,6,8)	12
Demand	4	4	6	8	8	

The minimum fuzzy transportation cost is $\widetilde{Z} = (114, 70, 58, 102)$.

Example 3: Consider an unbalance uncertain transportation problem as shown in Table 4.8.

Table 4.8 Unbalanced uncertain transportation problem.

	W_1	W_2	W_3	Supply
F_1	(2,3,4,3)	(1,1,1,1)	(4,5,4,3)	300
F_2	(3,2,1,2)	(6,7,3,8)	(8,9,9,10)	500
F_3	(8,9,7,8)	(5,2,3,2)	(3,2,1,2)	400
Demand	250	350	200	

The minimum fuzzy transportation cost is $\widetilde{Z} = (2350, 1600, 1650, 1600)$.

4.6 Result Comparison

We solved the three transportation problems (two balanced and one unbalanced) to illustrate the proposed approach. Also we solved the same examples with some existing methods [2, 18, 28, 30]. As the solution obtained is trapezoidal fuzzy number, to compare the results, we are using the ranking function. Table 4.9 presents a comparison of the acquired results.

Table 4.9 Result comparison.

Transportation problem	Balanced/ unbalanced	Gani and Razak [2]	Ritha and Merline [30]	Sudhakar and Kumar [28]	Hashmi et al. [18]	Proposed approach
Example 1	Balanced	168	166	168	162	160
Example 2	Balanced	92	90	90	86	86
Example 3	Unbalanced	2900	2500	2700	1900	1800

4.7 Conclusion

In this chapter, a new approach is suggested to solve the problem of transportation when sources or destinations have the limitation of capacity. Three examples are illustrated for the proposed approach and also solved these examples with the existing methods. The result comparison highlights the effectiveness and benefits of the suggested approach.

References

[1] A. Noora, and P. Karami, "Ranking functions and its applications to fuzzy DEA," *International Mathematical Forum*, vol. 3, no. 30, pp. 1469-1480, december 2008.

[2] A. N. Gani, and K. A. Razak, "Two Stage Fuzzy Transportation Problem," *Journal of Physical Sciences*, vol.10, pp. 63–69, january 2006.

[3] B. Choudhary, "Optimal solution of transportation problem based on revised distribution method," *International Journal of Innovative Research in Science Engineering and Technology*, vol. 5, no. 8, pp. 254-257, August 2016.

[4] E. E. Hosseini, "Three new methods to find initial basic feasible solution of transportation problems," *Applied Mathematical Sciences*, vol. 11, no. 37, pp. 1803-1814,July 2017.

[5] F. L Hitchcock, "The distribution of a product from several sources to numerous localities," *Journal of Mathematics and Physics*, vol. 20, no. 1-4, pp. 224–230, April 1941.

[6] G. Gupta, D. Rani and S. Singh, "An alternate for the initial basic feasible solution of category 1 uncertain transportation problems," *Proceedings of the National Academy of Sciences, India Section A: Physical Sciences*, vol. 90, March 2020, pp. 157-167,

[7] G. Gupta, S. Singh, and D. Rani, "A note on zero suffix method for the optimal solution of the transportation problems," *National Academy Science Letters*, vol. 41, pp. 293-294, August 2018,

[8] G. Singh and A. Singh, "Soft computing approach for three-level time minimization transportation problem," *International Conference on Technological Advancements in Computational Sciences, IEEE*, December 2022, pp. 236-24.

[9] H. A. Hussein and M. A. K. Shiker, "A modification to Vogel's approximation method to solve transportation problems," *Journal of Physics: Conference Series – IOP Publishing*, vol. 1591, no.1,June 2020, pp. 012029.

[10] H. I. Calvete, C. Gale, J. A. Iranzo, and P. Toth, "A metaheuristic for the two-stage fixed-charge transportation problem,"*Computers & Operations Research*, vol. 95, pp. 113-122, July 2018.

[11] J. Singla, G. Gupta, M. K. Kakkar and N. Garg, "A novel approach to find initial basic feasible solution of transportation problems under uncertain environment," *AIP Conference Proceedings*, vol. 2357, May 2022, pp. 110007.

[12] J. Singla, G. Gupta, M. K. Kakkar and N. Garg, "Revised algorithm of Vogel's approximation method (RA-VAM): an approach to find basic initial feasible solution of transportation problem," *ECS Transactions*, vol. 107, no. 1, April 2022, pp. 8757.

[13] K. Karagul and Y. Sahin, "A novel approximation method to obtain initial basic feasible solution of transportation problem," *Journal of King Saud University-Engineering Sciences*, vol. 32, no. 3, pp. 211-218, March 2020.

[14] L. A. Zadeh, "Fuzzy sets," *Information and Control*, vol. 8, no. 3, pp. 338–353, June 1965.

[15] M. M. Hossain, M. M. Ahmed, M. A. Islam and S. I. Ukil, "An effective approach to determine an initial basic feasible solution: A TOCM-MEDM Approach," *Open Journal of Optimization*, vol. 9, no. 2, pp. 27-37, June 2020.

[16] M. Malireddy, "A New Algorithm for initial basic feasible solution of Transportation Problem," *International Journal of Engineering Science Invention*, vol. 7, no. 8, pp. 41–43, August 2018.

[17] M. Sathyavathy and M. Shalini, "Solving transportation problem with four different proposed mean method and comparison with existing methods for optimum solution," *Journal of Physics: Conference Series, IOP Publishing*, vol. 1362, no. 1, June 2019, pp. 012088.

[18] N. Hashmi, S.A. Jalil and S. Javaid, "A model for two-stage fixed charge transportation problem with multiple objectives and fuzzy linguistic preferences," *Soft Computing*, vol. 23, pp. 12401-12415, January 2019.

[19] O. Mutlu, K. Karagül, & Y. Şahin,, "Avoid maximum cost method for determining the initial basic feasible solution of the transportation problem," *Pamukkale Üniversitesi Mühendislik Bilimleri Dergisi,* vol. 28, no. 4, pp. 569-576, August 2022.

[20] P. Pandian and G. Natarajan, "A new method for finding an optimal solution for transportation problems," *International Journal of Mathematical Sciences & Engineering Applications*, vol. 4, no. 2, pp. 59–65, January 2010.

[21] R. Kumar, R. Gupta and O. Karthiyayini, "A new approach to find the initial basic feasible solution of a transportation problem," *International Journal of Research – Granthaalayah*, vol. 6, no. 5, pp. 321-325, May 2018.

[22] R. Singh, R. Kaur and A. Singla, "Two phase fuzzy based prediction model to predict Soil nutrient," *Fifth International Conference on Parallel Distributed and Grid Computing, IEEE*, December 2018, pp. 80-85.

[23] S. Jamali, A. S. Soomro and M. M Shaikh, "The minimum demand method–a new and efficient initial basic feasible solution method for transportation problems," *Journal of Mechanics of Continua and Mathematical Sciences*, vol. 15, no. 19, pp. 94-109, September 2020.

[24] S. Kirupavathi, and M. Arthi, "Fuzzy transportation problem methodology using trapezoidal fuzzy numbers to selection the best treatment for blood cancer," *European Chemical Bulletin,* vol. 12, no. 7, pp. 3601-3613, 2023.

[25] Sonia and R. Malhotra, "A polynomial algorithm for a two – stage time minimizing transportation problem," Opsearch, vol. 39, no.5- 6, pp.251 – 266, December 2002.

[26] T. Can and H. Koçak, "Tuncay can's approximation method to obtain initial basic feasible solution to transport problem," *Applied and Computational Mathematics*, vol. 5, no. 2, pp. 78-82, May 2016.

[27] T.C. Koopmans, "Optimum utilization of the transportation system," *Econometrica*, vol. 17, pp. 136–146, july 1949.

[28] V. J. Sudhakar and V. N. Kumar, "A different approach for solving two stage fuzzy transportation problems,"*International Journal of Contemporary Mathematical Sciences*, vol. 6, no. 11, pp. 517–526, 2011.

[29] V. Jaiswal, A. K. Agrawal and A. Pandey, "Solving fuzzy transportation problem by various methods and their comparison in fuzzy environment," *European Chemical Bulletin*, vol. 12, pp. 2079-2084, may 2023.

[30] W. Ritha and J.M. Vinotha, "Multiobjective two stage fuzzy transportation problem,"*Journal of Physical Sciences*, vol. 13, pp.107–120, 2009.

5

A Note on Existence and Controllability Results on Fractional Sobolev-type Differential System of Order $\alpha \in (0, 1)$ in a Banach Space

Ankit Kumar[1], T. R. Mudgal[2], Neetu Ram[3], and Vandana Yadav[4]

[1]Department of Mathematics
Graphic Era Hill University, India
[2]Department of Applied Sciences & Humanities
State Institute of Engineering & Technology, India
[3]Department of Mathematics, Harsh Vidhya Mandir PG Collage Raisi
India
[4]Department of Mathematics & Scientific Computing
National Institute of Technology, India
E-mail: ankitramkumar620@gmail.com; trmudgal@gmail.com;
dr.n.chaudhary.rke@gmail.com; vandanarao1229@gmail.com

Abstract

This chapter aims to investigate the exact controllability of a Sobolev-type differential equation involving Hilfer fractional derivative in a Banach space. To establish exact controllability results for the proposed control problem, the theory of propagation family $\{\mathscr{P}(t),\ t \geq 0\}$ generated by the pair $(\mathcal{A}, \mathcal{C})$ of operators $\mathcal{A}$ and $\mathcal{C}$ (we also do not assume that the operator $\mathcal{C}$ has necessarily a bounded inverse), theory of fractional calculus, measure of noncompactness, and Sadovskii's fixed point theorem are used. To illustrate our main results, we provide an example.

Keywords: Exact controllability, Sobolev-type, propagation family, measure of noncompactness, Sadovskii's fixed point theorem

5.1 Introduction

Various dynamical systems can be steered from an initial state to any arbitrary final state with the help of some admissible controls. Such dynamical systems are called controllable. Basically, controllability is the elementary concept in the mathematical control theory and it plays an important role in the development of modern mathematical control theory. It is now extensively used not only in control theory but also in numerous fields of engineering and science such as quantum system theory, control of electric bulk power systems, reactor control, chemical process control, aerospace engineering, etc. There are some important controllability results concerning semi-linear or nonlinear differential systems obtained by several authors [2, 7, 14, 15, 17, 18, 19, 20, 21, 26, 29, 30, 34] in infinite dimensional space using semigroup theory and some suitable fixed point theorems.

On the other hand, fractional calculus has gained considerable interest of research community in recent years [12, 16, 17, 22, 27, 28, 31]. Fractional calculus is the generalization of the classical calculus from integer order derivative to arbitrary order. Now it is widely used in several areas of science and engineering due to its immense application such as signal and image processing, aerodynamics and biophysics, etc. There are several types of fractional integral and differential operators that came into existence. Riemann–Liouville and Caputo fractional derivatives are one of the most popular of them [16, 28]. At the end of the twentieth century, a new definition of fractional differential operator introduced by R. Hilfer [13] when he studied fractional time evolution in physical phenomena. This fractional derivative is known as Hilfer fractional derivative of order $0 < \alpha < 1$ and a type $0 \leq \nu \leq 1$. Hilfer fractional derivative becomes the Riemann–Liouville and Caputo fractional derivatives when $\nu = 0$ and $\nu = 1$, respectively. Hilfer fractional derivative appears in the theoretical simulation of dielectric relaxation in glass-forming materials, rheological constitute modeling, polymer science, etc. [5, 8, 9, 10, 33]. Moreover, the Sobolev-type differential equations frequently appear in the mathematical modeling of many physical phenomena such as in thermodynamics, the flow of fluid through fissured rocks, soil mechanics, shear in second-order fluids, etc., and it has been extensively studied by several authors [1, 2, 3, 4, 9, 26]. In [32], Wang et. al. established the controllability results for fractional evolution systems of Sobolev-type by considering Caputo fractional derivative and using the fixed point approach, measure of noncompactness, and propagation family.

Recently, Gou and Li [9] studied the nonlinear Sobolev-type delay differential equations with Hilfer fractional derivative using propagation family measure of non compactness condition. However, to the best of our knowledge, the exact controllability results of Sobolev-type fractional differential equations with Hilfer fractional derivative using propagation family and measure of non compactness condition have not been explored. Inspired by the above facts, we consider the following:

$$\begin{cases} D^{\alpha,\nu}\mathcal{C}\upsilon(t) = \mathcal{A}\upsilon(t) + \mathcal{C}f\left(t, \upsilon(t)\right) + \mathcal{C}\mathcal{B}\,\omega(t), & t \in \mathcal{I} = [0, \mathcal{T} \\ \mathcal{J}^{(1-\alpha)(1-\nu)}\upsilon(0) = X\mathcal{C}\upsilon_0, \end{cases} \tag{5.1}$$

where the Hilfer fractional derivative of order $0 < \alpha < 1$ and type $0 \le \nu \le 1$ is denoted by $D^{\alpha,\nu}$; the state function $\upsilon(\cdot)$ takes the value in Banach space $\mathcal{V}$; $\mathcal{J}^{\gamma}$ is a Riemann–Liouville fractional integral of order $\gamma > 0$; $\mathcal{A}$ and $\mathcal{C}$ are closed linear operators with domains contained in $\mathcal{V}$; control function $\omega(\cdot) \in L^2(\mathcal{I}, \mathcal{W})$, a Banach space of all admissible control functions, $\mathcal{W}$ is another Banach space of admissible control functions; $\mathcal{B}$ is a bounded linear operator from $\mathcal{W}$ into $\mathcal{W}$; the nonlinear function $f\colon \mathcal{I} \times \mathcal{V} \to \mathcal{V}$ is to be specified later.

To investigate the exact controllability results for system (5.1) we use the propagation family $\{\mathscr{P}(t),\ t \ge 0\}$ generated by the pair $(\mathcal{A}, \mathcal{C})$ of operators $\mathcal{A}$ and $\mathcal{C}$. To simplify the process, we construct $\{\mathscr{S}_{\alpha,\nu}(t),\ t \ge 0\}$ and $\{\mathscr{T}_{\alpha}(t),\ t \ge 0\}$ associated with the family of strongly continuous operators $\{\mathscr{P}(t),\ t \ge 0\}$. Also, we considered that the operator $\mathcal{C}$ is not necessarily a bounded inverse and the closed linear operators $\mathcal{A}$ and $\mathcal{C}$ are not are inevitably bounded. The theory of measure of noncompactness and Sadovskii's fixed point theorem are also used in our analysis.

The rest of the chapter is arranged as follows. In Section 5.2, we introduce some notations, basic definitions, and preliminary results. In Section 5.3, we prove the exact controllability of differential system (5.1) of Sobolev-type. In Section 5.4, we prove the controllability results for integro-differential system with nonlocal condition. Finally, in Section 5.5, we provide an example to show the application of our main results.

5.2 Preliminaries

Let $\mathcal{I}' = (0, \mathcal{T}]$ and $\hat{\mathcal{V}} = \{\upsilon \in C(\mathcal{I}', \mathcal{V}) : \lim_{t\to 0} t^{(1-\nu)(1-\alpha)}$ $\upsilon(t)$ exists and finite$\}$. Obviously $\hat{\mathcal{V}}$ is a Banach space with norm $\|\cdot\|_{\hat{\mathcal{V}}}$ defined by $\|\upsilon\|_{\hat{\mathcal{V}}} = \sup_{t\in\mathcal{I}'} \|t^{(1-\nu)(1-\alpha)}\upsilon(t)\|$.

Definition 5.1 ([13]). The Hilfer fractional derivative of the type $0 \leq \nu \leq 1$ and of order $0 < \alpha < 1$ with lower limit 0 for a function f is defined as

$$D_{0+}^{\alpha,\nu} f(t) = \left(I_{0+}^{\nu(1-\alpha)} \frac{d}{dt} (I_{0+}^{(1-\mu)} f) \right)(t) = \left(I_{0+}^{\nu(1-\alpha)} (D^{\mu} f) \right)(t),$$

where $\mu = \alpha + \nu - \alpha\nu$, provided that the right–hand side exists.

Remark 5.1. If $\nu = 0$, $0 < \alpha < 1$, then the Hilfer fractional derivative becomes the Riemann–Liouville fractional derivative of order α.

Consider the Cauchy problem as follows:

$$\begin{cases} \frac{d}{dt} \mathcal{C} v(t) = \mathcal{A} v(t), & t \in \mathcal{I}, \\ \mathcal{C} v(0) = \mathcal{C} v_0. \end{cases} \tag{5.2}$$

Definition 5.2 (see [24]). A family $\{\mathscr{P}(t)\}_{t\geq 0}$ of strongly continuous operators from $D(\mathcal{C})$ to $\mathcal{V}$, with constants $c > 0$, $\tilde{\mathcal{M}} > 0$ such that

$$\|\mathscr{P}(t)v\| \leq \tilde{\mathcal{M}} e^{ct} \|v\|, \quad t \geq 0 \text{ and } v \in D(\mathcal{C}),$$

is an exponentially bounded propagation family for system (5.2) if

$$(\lambda \mathcal{C} - \mathcal{A})^{-1} \mathcal{C} v = \int_0^{\infty} e^{-\lambda t} \mathscr{P}(t) v \, dt, \quad \text{for } \lambda > c \text{ and } v \in D(\mathcal{C}). \tag{5.3}$$

The pair $(\mathcal{A}, \mathcal{C})$ is said to be the generator of propagation family $\{\mathscr{P}(t)\}_{t\geq 0}$ if eqn (5.3) holds.

Definition 5.3 ([9, 10]). A continuous function $v\colon [0, \mathcal{T}] \to \mathcal{V}$ is said to be a mild solution of system (5.1) if the following fractional integral equation is satisfied:

$$v(t) = \mathscr{S}_{\alpha,\nu}(t) v_0 + \int_0^t \mathscr{T}_{\alpha}(t - \varphi) \left[f\left(\varphi, v(\varphi)\right) + \mathcal{B} \omega(\varphi) \right] d\varphi, \tag{5.4}$$

where

$$\begin{aligned} \mathscr{S}_{\alpha,\nu}(t) u &= I^{\nu(1-\alpha)} \mathscr{T}_{\alpha}(t) u, \quad \mathscr{T}_{\alpha}(t) u = t^{\alpha-1} Q_{\alpha}(t) u, \\ Q_{\alpha}(t) u &= \alpha \int_0^{\infty} \vartheta \psi_{\alpha}(\vartheta) \mathscr{P}(t^{\alpha} \vartheta) u \, d\vartheta, \\ \psi_{\alpha}(\vartheta) &= \frac{1}{\alpha} \vartheta^{-1-1/\alpha} \xi_{\alpha}(\vartheta^{-1/\alpha}), \\ \xi_{\alpha}(\vartheta) &= \frac{1}{\pi} \sum_{n=1}^{\infty} (-1)^{n-1} \vartheta^{-n\alpha-1} \frac{\Gamma(n\alpha + 1)}{n!} \sin(n\pi\alpha), \\ & \quad \vartheta \in (0, \infty), \quad u \in D(\mathcal{C}). \end{aligned}$$

Lemma 5.1 ([10]). If $\{\mathscr{P}(t)\}_{t\geq 0}$ is norm continuous and $\|\mathscr{P}(t)\| \leq \tilde{\mathcal{M}}$ for some $\tilde{\mathcal{M}} \geq 1$ and any $t > 0$, then $\mathscr{T}_\alpha(t)_{t>0}$ and $\mathscr{S}_{\nu,\alpha}(t)_{t>0}$ are strongly continuous linear operators, and for any $v \in \mathcal{V}$ and $t > 0$, we have

$$\|\mathscr{T}_\alpha(t)v\| \leq \frac{\tilde{\mathcal{M}} t^{\alpha-1}}{\Gamma(\alpha)}\|v\| \quad \text{and} \quad \|\mathscr{S}_{\nu,\alpha}(t)v\| \leq \frac{\tilde{\mathcal{M}} t^{(\nu-1)(1-\alpha)}}{\Gamma(\nu(1-\alpha)+\alpha)}\|v\|.$$

Definition 5.4 (see [6, 11]). Let $\mathscr{B}(\mathcal{V})$ be a collection of bounded subset of $\mathcal{V}$. A function $\chi\colon \mathscr{B}(\mathcal{V}) \to \mathbb{R}^+$ defined by

$$\begin{aligned}\chi(\mathcal{D}) &= \inf\{\varepsilon > 0 : \mathcal{D} \subset \bigcup_{j=1}^{n} \mathcal{D}_j,\ diam(\mathcal{D}_j) < \varepsilon\,(j = 1, 2, \ldots, n \in \mathbb{N})\},\\ &\qquad \mathcal{D} \in \mathscr{B}(\mathcal{V}),\end{aligned}$$

is called the Kuratowski's measure of noncompactness.

Lemma 5.2 (see [6, 11]). If G_1, G_2, and G are bounded subsets of a Banach space $\mathcal{V}$, then the following conditions hold:

(i) $\chi(G) = 0$ if and only if G is a relatively compact set in $\mathcal{V}$.
(ii) $\chi(G_1) \leq \chi(G_2)$ if $G_1 \subset G_2$.
(iii) $\chi(G_1 + G_2) \leq \chi(G_1) + \chi(G_2)$.
(iv) $\chi(cG) \leq |c|\chi(G)$ for any $c \in \mathbb{R}$.

Lemma 5.3 (see [23]). If $S \subset \mathcal{V}$ is bounded, there is a countable set $G \subset S$ such that $\chi(S) \leq 2\chi(G)$.

Definition 5.5. Let X and Y be two Banach spaces. A continuous map $F\colon \Omega \subset X \to Y$ is said to be a condensing map if F takes the bounded sets into bounded sets and $\chi(F(S)) < \chi(S)$ for all bounded sets $S \subset \Omega$ with $\chi(S) \neq 0$.

Theorem 5.1 (Sadovskii's fixed point theorem). Let X be a Banach space and Ω be a closed convex bounded subset in X. If $F : \Omega \to \Omega$ is a condensing map, then F has a fixed point in Ω.

5.3 Controllability Results

In this section, we discuss the existence of the mild solution and establish a set of controllability results for the Hilfer fractional differential system (5.1) of Sobolev-type in a Banach space using Sadovskii's fixed point theorem.

Definition 5.6. System (5.1) is said to be controllable on the interval $\mathcal{I}$ if for every $\upsilon_0, \upsilon_{\mathcal{T}} \in \mathcal{V}$, there exists a control $\omega(\cdot) \in L^2(\mathcal{I}, \mathcal{W})$ such that the mild solution $\upsilon(\cdot)$ of system (5.1) satisfies $\upsilon(\mathcal{T}) = \upsilon_{\mathcal{T}}$.

In order to establish the results, we require the following hypothesis:

(H0) $\{\mathscr{P}(t)\}_{t\geq 0}$ is norm continuous and uniformly bounded.

(H1) The function $f : \mathcal{I} \times \mathcal{V} \to \mathcal{V}$ satisfies the Caratheodary condition, that is, the function $f(t, \cdot) : \mathcal{V} \to \mathcal{V}$ is continuous for every $t \in \mathcal{I}$ and $f(\cdot, \upsilon) \colon \mathcal{I} \to \mathcal{V}$ is strongly measurable for any $\upsilon \in \mathcal{V}$. For each $t \in \mathcal{I}$ and $\epsilon > 0$, there exists $\kappa(t) \in L^1([0, t], \mathbb{R}^+)$ such that

$$\sup\{\|f(t,\upsilon)\| \colon \|t^{(1-\nu)(1-\alpha)}\upsilon\| \leq \epsilon\} \leq \kappa(t) \quad \text{a.e. for } t \in \mathcal{I},$$

$$\text{and} \quad \liminf_{\epsilon \to \infty} \frac{1}{\epsilon} \int_0^t (t-\varphi)^{1-\alpha} \kappa(\varphi) d\varphi = l < +\infty.$$

(H2) There exists a function $\eta \colon (\mathcal{I} = [0, \mathcal{T}]) \to \mathbb{R}^+$ with the condition $(t - \cdot)^{1-\alpha}\eta(\cdot) \in L^1([0, t], \mathbb{R}^+)$ such that

$$\chi(f(t, G)) \leq \eta(t)\chi(t^{(1-\nu)(1-\alpha)}G)$$

for $t \in \mathcal{I}$, and $G \subset \mathcal{V}$. Also suppose that $\eta^* = \max_{t\in\mathcal{I}} \int_0^t (t - \varphi)^{1-\alpha}\eta(\varphi)\, d\varphi$.

Now we define the linear operator $\Upsilon \colon L^2(\mathcal{I}, \mathcal{W}) \to \mathcal{V}$ as follows:

$$\Upsilon\omega = \int_0^{\mathcal{T}} \mathscr{T}_\alpha(\mathcal{T} - \varphi)\mathcal{B}\omega(\varphi)d\varphi$$

has an inverse operator Υ^{-1} that takes values in $L^2(\mathcal{I}, \mathcal{W})/\ker T$ and there exists a constant $\mathcal{N}_\Upsilon > 0$ such that

$$\|\Upsilon^{-1}\|_{L^2} \leq \mathcal{N}_\Upsilon,$$

and also there is an integrable function $K_w \colon \mathcal{I} \to \mathbb{R}^+$ such that for every bounded set $G \subset \mathcal{V}$,

$$\chi(\Upsilon^{-1}G(t)) \leq K_w(t)\chi(G) \quad \text{and} \quad K_w^* = \max_{t\in\mathcal{I}} \int_0^t (t-\varphi)^{1-\alpha} K_w(\varphi)\, d\varphi.$$

Also, we take

$$N = \frac{2\tilde{\mathcal{M}}\mathcal{T}^{(1-\nu)(1-\alpha)}}{\Gamma(\alpha)}\eta^* \left(1 + \frac{\tilde{\mathcal{M}}\|\mathcal{B}\|\, K_w^*}{\Gamma(\alpha)}\right) < \frac{1}{2}.$$

Now we define the sets $\mathcal{S}_\epsilon$ and $\mathcal{S}_\epsilon^{\hat{\mathcal{V}}}$ as

$$\mathcal{S}_\epsilon = \{u(\cdot) \in C(\mathcal{I}, \mathcal{V}) \colon \|u\|_1 \le \epsilon\} \quad \text{and} \quad \mathcal{S}_\epsilon^{\hat{\mathcal{V}}} = \{\upsilon \in \hat{\mathcal{V}} \colon \|\upsilon\|_{\hat{\mathcal{V}}} \le \epsilon\}.$$

Theorem 5.2. Assume that hypothes (H0)–(H2) hold. Then system (5.1) is controllable on $\mathcal{I}$ if the condition

$$\Theta\left(1 + \frac{\tilde{\mathcal{M}}\mathcal{T}^\alpha}{\Gamma(\alpha+1)}\|\mathcal{B}\|\mathcal{N}_\Upsilon\right) < 1 \tag{5.5}$$

is satisfied.

Proof. We define the control function $\hat{\omega}$ for the control system (5.1), $\upsilon(\cdot) \in C(\mathcal{I}, \mathcal{V})$, as follows:

$$\hat{\omega}(t) = \Upsilon^{-1}\left[\upsilon_{\mathcal{T}} - \mathscr{S}_{\alpha,\nu}(\mathcal{T})\upsilon_0 - \int_0^{\mathcal{T}} \mathscr{T}_\alpha(\mathcal{T}-\varphi) f(\varphi, \upsilon(\varphi))\, d\varphi\right](t), \tag{5.6}$$

Using the control system (5.6), we define an operator $\mathcal{F} \colon \hat{\mathcal{V}} \to \hat{\mathcal{V}}$ as

$$(\mathcal{F}\upsilon)(t) = \mathscr{S}_{\alpha,\nu}(t)\upsilon_0 + \int_0^t \mathscr{T}_\alpha(t-\varphi)\left[f(\varphi, \upsilon(\varphi)) + \mathcal{B}\hat{\omega}(\varphi)\right] d\varphi. \tag{5.7}$$

For any $u \in C(\mathcal{I}, \mathcal{V})$, $\upsilon(t) = t^{(\nu-1)(1-\alpha)}u(t)$, $t \in \mathcal{I}'$, we define a map $\mathcal{Q}$ as

$$(\mathcal{Q}u)(t) = \begin{cases} t^{(1-\nu)(1-\alpha)}(\mathcal{F}\upsilon)(t), & t \in (0, \mathcal{T}], \\ \frac{\upsilon_0}{\Gamma[\nu(1-\alpha)+\alpha]}, & t = 0. \end{cases} \tag{5.8}$$

It is obvious that u is any fixed point of map $\mathcal{Q}$ if and only if υ is a fixed point of $\mathcal{F}$, which is further equivalent to say that $\upsilon(\cdot)$ is the mild solution of eqn (5.1). If $\upsilon(\cdot)$ is the mild solution of eqn (5.1) with the control system (5.6), then $\upsilon(\mathcal{T}) = \upsilon_{\mathcal{T}}$.

Let $u \in \mathcal{S}_\epsilon$ and $\upsilon(t) = t^{(\nu-1)(1-\alpha)}u(t)$, $t \in \mathcal{I}'$. Therefore, $\upsilon \in \mathcal{S}_\epsilon^{\hat{\mathcal{V}}}$. We now obtain from eqn (5.6) and Lemma 5.1 that

$$\|\mathcal{B}\hat{\omega}(t)\| \le \|\mathcal{B}\|\mathcal{N}_\Upsilon\left[\|\upsilon_{\mathcal{T}}\| + \frac{\tilde{\mathcal{M}}}{\Gamma(\nu(1-\alpha)+\alpha)}\mathcal{T}^{(\nu-1)(1-\alpha)}\upsilon_0\right.$$

$$
\begin{aligned}
&+\frac{\tilde{\mathcal{M}}}{\Gamma(\alpha)}\int_0^{\mathcal{T}}(\mathcal{T}-\varphi)^{\alpha-1}\left\|f\left(\varphi,v(\varphi)\right)\right\|d\varphi\Bigg] \\
&\leq \|\mathcal{B}\|\mathcal{N}_\Upsilon\left[\|v_{\mathcal{T}}\|+\tilde{M}v_0+\frac{\tilde{\mathcal{M}}}{\Gamma(\alpha)}\int_0^{\mathcal{T}}(\mathcal{T}-\varphi)^{\alpha-1}\kappa(\varphi)d\varphi\right] \\
&= \mathcal{N}^* \quad \text{(say)}. \qquad (5.9)
\end{aligned}
$$

First we would like to show that $\mathcal{Q}(\mathcal{S}_\epsilon) \subseteq \mathcal{S}_\epsilon$ for some $\epsilon > 0$. On the contrary, we suppose that this is not true for each $\epsilon > 0$. Thus, there exists $u_\epsilon \in \mathcal{S}_\epsilon$ for each $\epsilon > 0$ such that $\|\mathcal{Q}(u_\epsilon)(t_\epsilon)\| > \epsilon$ for some $t_\epsilon \in \mathcal{I}$. Let $v_\epsilon(t) = t^{(\nu-1)(1-\alpha)}u_\epsilon(t)$, $t \in \mathcal{I}'$. We obtain

$$
\begin{aligned}
\epsilon < \|\mathcal{Q}u_\epsilon(t_\epsilon)\| \leq\ & \frac{\tilde{\mathcal{M}}}{\Gamma(\nu(1-\alpha)+\alpha)}v_0+\frac{\tilde{\mathcal{M}}t_\epsilon^{(1-\nu)(1-\alpha)}}{\Gamma(\alpha)} \\
& \int_0^{t_\epsilon}(t_\epsilon-\varphi)^{\alpha-1}\left\|f\left(\varphi,v(\varphi)\right)+Bw(\varphi)\right\|d\varphi \\
\leq\ & \frac{\tilde{\mathcal{M}}}{\Gamma(\nu(1-\alpha)+\alpha)}v_0+\frac{\tilde{\mathcal{M}}t_\epsilon^{(1-\nu)(1-\alpha)}}{\Gamma(\alpha)} \\
& \int_0^{t_\epsilon}(t_\epsilon-\varphi)^{\alpha-1}\kappa(\varphi)d\varphi \\
& +\frac{\tilde{\mathcal{M}}t_\epsilon^{(1-\nu)(1-\alpha)}}{\Gamma(\alpha)}\int_0^{t_\epsilon}(t_\epsilon-\varphi)^{\alpha-1}\|\mathcal{B}w(\varphi)\|d\varphi. \qquad (5.10)
\end{aligned}
$$

Eqn (5.10) is divided by ϵ on both sides and then taking $\epsilon \to \infty$, we obtain

$$
1 < \Theta\left(1+\frac{\tilde{\mathcal{M}}\mathcal{T}^\alpha}{\Gamma(\alpha+1)}\|\mathcal{B}\|\mathcal{N}_\Upsilon\right),
$$

where $\Theta = \frac{\tilde{\mathcal{M}}l\mathcal{T}^{(1-\nu)(1-\alpha)}}{\Gamma(\alpha)}$.

This is a contradiction of our assumption that

$$
\Theta\left(1+\frac{\tilde{\mathcal{M}}\mathcal{T}^\alpha}{\Gamma(\alpha+1)}\|\mathcal{B}\|\mathcal{N}_\Upsilon\right) < 1.
$$

Thus, there is a $\epsilon > 0$ such that $\mathcal{Q}(\mathcal{S}_\epsilon) \subset \mathcal{S}_\epsilon$.

We shall now show that $\mathcal{Q}$ is continuous on $\mathcal{S}_\epsilon$. Let $\{u_n\} \subset \mathcal{S}_\epsilon$ with $u_n \to u \in \mathcal{S}_\epsilon$ as $n \to \infty$. Take $v_n(t) = t^{(\nu-1)(1-\alpha)}u_n(t)$, $t \in \mathcal{I}'$. Then we have the following:

1. $\left\| f(t, v_n(t)) - f(t, v(t)) \right\| \leq 2\kappa(t)$.
2. $\left\| f(t, v_n(t)) - f(t, v(t)) \right\| \to 0$ as $n \to \infty$.
3. $\|\mathcal{B}\hat{\omega}_n(t) - \mathcal{B}\hat{\omega}(t)\| \leq 2\tilde{\mathcal{M}}^*$.

Using the above conditions 1, 2, 3 with Lebesgue dominated convergence theorem, for $t \in \mathcal{I}'$, we get

$$\begin{aligned}
&\|\mathcal{B}\hat{\omega}_n(t)) - \mathcal{B}\hat{\omega}(t)\| \leq \|\mathcal{B}\| \, \mathcal{N}_\Upsilon \\
&\left[\frac{\tilde{\mathcal{M}}}{\Gamma(\alpha)} \int_0^{\mathcal{T}} (\mathcal{T} - \varphi)^{\alpha-1} \left\| f(\varphi, v_n(\varphi)) - f(\varphi, v(\varphi)) \right\| d\varphi \right] \\
&\to 0 \text{ as } n \to \infty.
\end{aligned}$$

Therefore, we obtain for $t \in \mathcal{I}'$ that

$$\begin{aligned}
\|\mathcal{Q}u_n(t)) - \mathcal{Q}u(t)\| \leq\ & t^{(1-\nu)(1-\alpha)} \frac{\tilde{\mathcal{M}}}{\Gamma(\alpha)} \\
& \int_0^t (t-\varphi)^{\alpha-1} \Big[\left\| f(\varphi, v_n(\varphi)) - f(\varphi, v(\varphi)) \right\| \\
& + \left\| \mathcal{B}\hat{\omega}_n(\varphi) - \mathcal{B}\hat{\omega}(\varphi) \right\| \Big] d\varphi \\
& \to 0 \text{ as } n \to \infty.
\end{aligned}$$

This shows that $\mathcal{F}$ is continuous on $\mathcal{S}_\epsilon$.

Next we shall prove that $\mathcal{F}(\mathcal{S}_\epsilon)$ is equicontinuous on $\mathcal{I}$. Let $u \in \mathcal{S}_\epsilon$ and $v(t) = t^{(\nu-1)(1-\alpha)}u(t)$, $t \in \mathcal{I}'$. For any $t_1, t_2 \in \mathcal{I}'$ with $t_1 < t_2$ and $v \in \mathcal{S}_\epsilon$, we have

$$\begin{aligned}
\|\mathcal{Q}u(t_2)) - \mathcal{Q}u(t_1)\| \leq\ & \left\| t_2^{(1-\nu)(1-\alpha)} \mathscr{S}_{\alpha,\nu}(t_2)v_0 - t_1^{(1-\nu)(1-\alpha)} \mathscr{S}_{\alpha,\nu}(t_1)v_0 \right\| \\
& + \left\| \int_0^{t_1} \left[t_2^{(1-\nu)(1-\alpha)} \mathscr{T}_\alpha(t_2 - \varphi) - t_1^{(1-\nu)(1-\alpha)} \mathscr{T}_\alpha(t_1 - \varphi) \right] \right. \\
& \left. [f(\varphi, v(\varphi)) + \mathcal{B}\hat{\omega}(\varphi)] \, d\varphi \right\| \\
& + \frac{\tilde{\mathcal{M}}}{\Gamma(\alpha)} t_2^{(1-\nu)(1-\alpha)} \int_{t_1}^{t_2} (t_2 - \varphi)^{\alpha-1} \times \left[\kappa(\varphi) + \mathcal{N}^* \right] d\varphi \\
& = \Lambda_1 + \Lambda_2 + \Lambda_3,
\end{aligned}$$

where

$$\Lambda_1 = \left\| t_2^{(1-\nu)(1-\alpha)} \mathscr{S}_{\alpha,\nu}(t_2)\upsilon_0 - t_1^{(1-\nu)(1-\alpha)} \mathscr{S}_{\alpha,\nu}(t_1)\upsilon_0 \right\|,$$

$$\Lambda_2 = \left\| \int_0^{t_1} \left[t_2^{(1-\nu)(1-\alpha)} \mathscr{T}_\alpha(t_2-\varphi) - t_1^{(1-\nu)(1-\alpha)} \mathscr{T}_\alpha(t_1-\varphi) \right] \right.$$
$$\left. [f(\varphi,\upsilon(\varphi)) + B\hat{\omega}(\varphi)]\, d\varphi \right\|,$$

$$\Lambda_3 = \frac{\tilde{\mathcal{M}}}{\Gamma(\alpha)} t_2^{(1-\nu)(1-\alpha)} \int_{t_1}^{t_2} (t_2-\varphi)^{\alpha-1} [\kappa(\varphi) + \mathcal{N}^*]\, d\varphi.$$

From the expression of Λ_3, we can easily see that Λ_3 tends to 0 as $t_2 \to t_1$ is independent of $\upsilon \in \mathcal{S}_\epsilon$.

$$\begin{aligned}
\Lambda_1 &= \frac{1}{\Gamma(\nu(1-\alpha))} \left\| t_2^{(1-\nu)(1-\alpha)} \int_0^{t_2} (t_2-\varphi)^{\nu(1-\alpha)-1} \varphi^{\alpha-1} Q_\alpha(\varphi)\upsilon_0\, d\varphi \right. \\
&\quad \left. - t_1^{(1-\nu)(1-\alpha)} \int_0^{t_1} (t_1-\varphi)^{\nu(1-\alpha)-1} \varphi^{\alpha-1} Q_\alpha(\varphi)\upsilon_0\, d\varphi \right\| \\
&\leq \frac{t_2^{(1-\nu)(1-\alpha)}}{\Gamma(\nu(1-\alpha))} \int_{t_1}^{t_2} (t_2-\varphi)^{\nu(1-\alpha)-1} \varphi^{\alpha-1} \|Q_\alpha(\varphi)\upsilon_0\|\, d\varphi \\
&\quad + \int_0^{t_1} \left| t_2^{(1-\nu)(1-\alpha)} (t_2-\varphi)^{\nu(1-\alpha)-1} \right. \\
&\quad \left. - t_1^{(1-\nu)(1-\alpha)} (t_1-\varphi)^{\nu(1-\alpha)-1} \right| \varphi^{\alpha-1} \|Q_\alpha(\varphi)\upsilon_0\|\, d\varphi \\
&\leq \frac{M t_1^{\alpha-1} t_2^{(1-\nu)(1-\alpha)}}{\Gamma(\nu(1-\alpha))\Gamma(\alpha)} \frac{\|\upsilon_0\|}{\nu(1-\alpha)} (t_2-t_1)^{\nu(1-\alpha)} + \frac{\daleth \|\upsilon_0\|}{\Gamma(\nu(1-\alpha))\Gamma(\alpha)} \\
&\quad \times \int_0^{t_1} \left| \left[t_2^{(1-\nu)(1-\alpha)} (t_2-\varphi)^{\nu(1-\alpha)-1} - t_1^{(1-\nu)(1-\alpha)} (t_1-\varphi)^{\nu(1-\alpha)-1} \right] \varphi^{\alpha-1} \right| d\varphi \\
&\quad \to 0 \text{ as } t_2 \to t_1 \text{ is independent of } \upsilon.
\end{aligned}$$

$$\begin{aligned}
\Lambda_2 &\leq \left\| \int_0^{t_1} \left[t_2^{(1-\nu)(1-\alpha)} (t_2-\varphi)^{\alpha-1} - t_1^{(1-\nu)(1-\alpha)} (t_1-\varphi)^{\alpha-1} \right] \right. \\
&\quad \left. Q_\alpha(t_2-\varphi) [f(\varphi,\upsilon(\varphi)) + B\hat{\omega}(\varphi)]\, d\varphi \right\| \\
&\quad + t_1^{(1-\nu)(1-\alpha)} \left\| \int_0^{t_1} (t_1-\varphi)^{\alpha-1} [Q_\alpha(t_2-\varphi) - Q_\alpha(t_1-\varphi)] [f(\varphi,\upsilon(\varphi)) \right. \\
&\quad \left. + B\hat{\omega}(\varphi)]\, d\varphi \right\| \\
&\leq \frac{\tilde{\mathcal{M}}}{\Gamma(\alpha)} \int_0^{t_1} \left\| t_2^{(1-\nu)(1-\alpha)} (t_2-\varphi)^{\alpha-1} - t_1^{(1-\nu)(1-\alpha)} (t_1-\varphi)^{\alpha-1} \right\| [\kappa(\varphi) + \mathcal{N}^*]\, d\varphi \\
&\quad + t_1^{(1-\nu)(1-\alpha)} \int_0^{t_1} (t_1-\varphi)^{\alpha-1} \|[Q_\alpha(t_2-\varphi) - Q_\alpha(t_1-\varphi)]\| \\
&\quad [\kappa(\varphi) + \mathcal{N}^*]\, d\varphi = \Lambda_{2,1} + \Lambda_{2,2},
\end{aligned}$$

where

$$\Lambda_{2,1} = \frac{\tilde{\mathcal{M}}}{\Gamma(\alpha)} \int_0^{t_1} \left| t_2^{(1-\nu)(1-\alpha)}(t_2-\varphi)^{\alpha-1} - t_1^{(1-\nu)(1-\alpha)}(t_1-\varphi)^{\alpha-1} \right| [\kappa(\varphi) + \mathcal{N}^*]\, d\varphi$$

and

$$\Lambda_{2,2} = t_1^{(1-\nu)(1-\alpha)} \int_0^{t_1} (t_1-\varphi)^{\alpha-1} \|[\mathcal{Q}_\alpha(t_2-\varphi) - \mathcal{Q}_\alpha(t_1-\varphi)] [\kappa(\varphi) + \mathcal{N}^*]\|\, d\varphi.$$

If $\delta \in (0, t_1)$, then we have

$$\begin{aligned}
\Lambda_{2,2} \le{}& t_1^{(1-\nu)(1-\alpha)} \int_0^{t_1-\delta} (t_1-\varphi)^{\alpha-1} \|\mathcal{Q}_\alpha(t_2-\varphi) - \mathcal{Q}_\alpha(t_1-\varphi)\| [\kappa(\varphi) + \mathcal{N}^*]\, d\varphi \\
&+ \frac{\tilde{\mathcal{M}} t_1^{(1-\nu)(1-\alpha)}}{\Gamma(\alpha)} \int_{t_1-\delta}^{t_1} (t_1-\varphi)^{\alpha-1} [\kappa(\varphi) + \mathcal{N}^*]\, d\varphi \\
\le{}& t_1^{(1-\nu)(1-\alpha)} \int_0^{t_1-\delta} (t_1-\varphi)^{\alpha-1} [\kappa(\varphi) + \mathcal{N}^*]\, d\varphi \\
&\times \sup_{\varphi \in [0, t-\delta]} \|\mathcal{Q}_\alpha(t_2-\varphi) - \mathcal{Q}_\alpha(t_1-\varphi)\| \\
&+ \frac{\tilde{\mathcal{M}} t_1^{(1-\nu)(1-\alpha)}}{\Gamma(\alpha)} \int_{t_1-\delta}^{t_1} (t_1-\varphi)^{\alpha-1} [\kappa(\varphi) + \mathcal{N}^*]\, d\varphi.
\end{aligned}$$

Since $(t-\varphi)^{\alpha-1}\kappa(\varphi)$ is integrable over the interval $[0, t]$, and the continuity of $\mathcal{Q}_\alpha(t)$ in uniform operator topology for $t > 0$, it is obvious that $\Lambda_{2,1}$ and $\Lambda_{2,2}$ tend to 0 as $t_2 \to t_1$ and $\delta \to 0$ is independent of u. Therefore, $\|\mathcal{Q}u(t_2)) - \mathcal{Q}u(t_1)\| \to 0$ as $t_2 \to t_1$ is independent of u. That is, $\mathcal{Q}(\mathcal{S}_\epsilon)$ is a family of equicontinuous functions on $\mathcal{I}'$. Since $t^{(1-\nu)(1-\alpha)}\mathscr{S}_{\alpha,\nu}(t)$ is uniformly continuous on $\mathcal{I}$, the family $\mathcal{Q}(\mathcal{S}_\epsilon)$ is equicontinuous on $\mathcal{I}$.

Let S be any subset of $\mathcal{S}_\epsilon$. From Lemma 5.3, we get a countable set $S_1 = \{u_n\} \subset S$ such that

$$\chi(\mathcal{Q}(S)) \le 2\chi(\mathcal{Q}(S_1)).$$

Let $v_n(t) = t^{(\nu-1)(1-\alpha)}u_n(t)$. Then we have

$$\begin{aligned}
\chi(\mathcal{Q}(S_1)(t)) &\leq \chi\left(\left\{t^{(1-\nu)(1-\alpha)}\mathscr{S}_{\alpha,\nu}(t)\upsilon_0)\right\}\right)\\
&\quad+\chi\left(\left\{t^{(1-\nu)(1-\alpha)}\int_0^t \mathscr{T}_\alpha(t-\varphi)f\left(\varphi,\upsilon_n(\varphi)\,d\varphi\right\}\right)\\
&\quad+\chi\left(\left\{t^{(1-\nu)(1-\alpha)}\int_0^t \mathscr{T}_\alpha(t-\varphi)B\hat{\omega}_n(\varphi)d\varphi\right\}\right)\\
&\leq \frac{2\tilde{\mathcal{M}}\mathcal{T}^{(1-\nu)(1-\alpha)}}{\Gamma(\alpha)}\int_0^t (t-\varphi)^{\alpha-1}\chi\left(\{f\left(\varphi,\upsilon_n(\varphi)\right)\}\right)d\varphi\\
&\quad+\frac{2\tilde{\mathcal{M}}\,\|\mathcal{B}\|\mathcal{T}^{(1-\nu)(1-\alpha)}}{\Gamma(\alpha)}\int_0^t (t-\varphi)^{\alpha-1}\\
&\quad\times K_w(\varphi)\left[\chi\left(\left\{\int_0^{\mathcal{T}} \mathscr{T}_\alpha(\mathcal{T}-\varphi)f\left(\varphi,\upsilon_n(\varphi)\right)d\varphi\right\}\right)\right]\\
&\leq \frac{2\tilde{\mathcal{M}}\mathcal{T}^{(1-\nu)(1-\alpha)}}{\Gamma(\alpha)}\int_0^t (t-\varphi)^{\alpha-1}\eta(\varphi)\\
&\quad\left[\chi\left(\left\{\varphi^{(1-\nu)(1-\alpha)}\upsilon_n(\varphi)\right\}\right)\right]d\varphi\\
&\quad+\frac{2\tilde{\mathcal{M}}\,\|\mathcal{B}\|\mathcal{T}^{(1-\nu)(1-\alpha)}}{\Gamma(\alpha)}\int_0^t (t-\varphi)^{\alpha-1}K_w(\varphi)\\
&\quad\left[\frac{\tilde{\mathcal{M}}}{\Gamma(\alpha)}\int_0^{\mathcal{T}}(\mathcal{T}-\varphi)^{\alpha-1}\chi\left\{f\left(\varphi,\upsilon_n(\varphi)\right)\right\}d\varphi\right]\\
&\leq \frac{2\tilde{\mathcal{M}}\,\mathcal{T}^{(1-\nu)(1-\alpha)}}{\Gamma(\alpha)}\eta^*\left(1+\frac{\tilde{\mathcal{M}}\|\mathcal{B}\|\,K_w^*}{\Gamma(\alpha)}\right)\sup_{0\leq\varphi\leq\mathcal{I}}\chi\left(\{u_n(\varphi)\}\right)\\
&\leq N\chi(\{u_n\}). \qquad (5.11)
\end{aligned}$$

Since $\mathcal{Q}(S_1) \subset \mathcal{Q}(\mathcal{S}_\epsilon)$ is equicontinuous, we have $\mu(\mathcal{Q}(S_1)) = \sup_{t\in\mathcal{I}}\mu(\mathcal{Q}(S_1)(t))$. Thus, from the equicontinuity of $\mathcal{Q}(\mathcal{S}_\epsilon)$ and hypothesis (A6), we have that

$$\chi(\mathcal{Q}(S)) \leq 2\chi(\mathcal{Q}(S_1)) \leq 2N\chi((S)).$$

Since $N < \frac{1}{2}$, the map is a condensing map from $\mathcal{S}_\epsilon$ to $\mathcal{S}_\epsilon$. Therefore, from Sadovskii's fixed point theorem, the map $\mathcal{Q}$ has a fixed point u in $\mathcal{S}_\epsilon$. Hence, the system has a mild solution $\upsilon(t) = t^{(\nu-1)(1-\alpha)}u(t)$ of system (5.1) satisfying that $\upsilon(\mathcal{T}) = \upsilon_{\mathcal{T}}$. That is, system (5.1) is exactly controllable on the interval $\mathcal{I} = [0,\mathcal{T}]$. This completes the proof. □

5.4 An Example

Let $\mathcal{V} = L^2([0,\pi],\mathbb{R})$ with L^2-norm $\|\cdot\|_2$. Consider the following fractional partial differential equations:

$$\begin{cases} D^{\alpha,\nu}\left[v(t,z) - v_{zz}(t,z)\right] = v_{zz}(t,z) + \Phi_1\left(t, v(t,z)\right) + \Phi_2(t,z), (t,z) \in [0,\mathcal{T}] \times [0,\pi], \\ v(t,0) = v(t,\pi) = 0, \ \ t \geq 0, \\ \mathcal{J}^{(1-\alpha)(1-\nu)}\left[v(t,z) - v_{zz}(t,z)\right] = \widehat{v}_0(z), \quad 0 \leq z \leq \pi, \end{cases} \tag{5.12}$$

where $D^{\alpha,\nu}$ is a Hilfer fractional partial derivative of order α and type ν, $0 \leq \nu \leq 1$, $0 < \alpha < 1$; $\widehat{v}_0 \in \mathcal{V}$ and $\mathcal{J}^{(1-\alpha)(1-\nu)}$ is the Riemann–Liouville fractional integral operator.

Define $\mathcal{A}: \ D(\mathcal{A}) \subset \mathcal{V} \to \mathcal{V}$ by

$$\mathcal{A}v = v_{zz} \tag{5.13}$$

and $\mathcal{C}: \ D(\mathcal{C}) \subset \mathcal{V} \to \mathcal{V}$ by $\mathcal{C}v = v - v_{zz}$, where $D(\mathcal{A}) = D(\mathcal{C}) = \{v \in \mathcal{V} : v, \ v_z$ are absolutely continuous, $v_{zz} \in \mathcal{V}, \ v(0) = v(\pi) = 0\}$. Then we can write $\mathcal{A}$ and $\mathcal{C}$ as follows [25]:

$$\mathcal{A}v = -\sum_{r=1}^{\infty} r^2 (v, \ v_r) v_r, \quad v \in D(\mathcal{A})$$

and

$$\mathcal{C}v = \sum_{r=1}^{\infty} (1 + r^2)(v, \ v_r) v_r, \quad v \in D(\mathcal{C})$$

where $v_r = \sqrt{2/\pi}\sin rz$, $r = 1, 2, \ldots$ is the orthonormal set of eigenfunctions of $\mathcal{A}$. Furthermore, for $v \in \mathcal{V}$, we have

$$\mathcal{C}^{-1}v = \sum_{r=1}^{\infty} \frac{1}{(1+r^2)}(v, \ v_r) v_r,$$

$$\mathcal{A}\mathcal{C}^{-1}v = \sum_{r=1}^{\infty} \frac{-r^2}{(1+r^2)}(v, \ v_r) v_r,$$

and the propagation family $\{\mathscr{P}(t), \ t \geq 0\}$ can be generated by $\mathcal{A}\mathcal{C}^{-1}$ and written as follows:

$$\mathscr{P}(t)v = \sum_{r=1}^{\infty} e^{\frac{-r^2}{(1+r^2)}t}(v, \ v_r) v_r.$$

It can be easily seen that $\{\mathscr{P}(t)$ is uniformly bounded and $\{\mathscr{P}(t),\, t \geq 0\}$ is norm continuous for $t > 0$ and $\|\mathscr{P}(t)\| \leq 1$ for all $0 \leq t \leq \mathcal{T}$. The bounded linear control operator $\mathcal{B} : L^2(\mathcal{I}, \mathcal{W}) \to L^2(\mathcal{I}, \mathcal{W})$ is defined by $\mathcal{B}\omega = \tilde{b}\omega$. The linear operator $\Upsilon : L^2(\mathcal{I}, \mathcal{W}) \to \mathcal{V}$ is given by

$$(\Upsilon\omega)(z) = \tilde{b} \int_0^{\mathcal{T}} \mathscr{T}_\alpha(\mathcal{T} - \varphi)\, \omega(\varphi, z) d\varphi$$

and has an inverse operator Υ^{-1} that takes values in $L^2(\mathcal{I}, \mathcal{W}) / \ker \Upsilon$, where $\ker \Upsilon = \{v \in (\mathcal{I}, \mathcal{W}) : \Upsilon v = 0\}$.

Now system (5.12) can be formulated abstractly as

$$\begin{cases} D^{\alpha,\nu}\mathcal{C}v(t) = \mathcal{A}v(t) + \mathcal{C}f\,(t, v(t)) + \mathcal{C}\mathcal{B}\,\omega(t), & t \in \mathcal{I} = [0, \mathcal{T}] \\ \mathcal{J}^{(1-\alpha)(1-\nu)}v(0) = \mathcal{C}v_0, \end{cases}$$

where $\Phi_1\,(t, v(t,z)) = \mathcal{C}f\,(t, v(t))$, $\Phi_2(t,z) = \omega(t)(z)$. The operator $\mathcal{A}$ is the same as in eqn (5.13). Thus, Theorem 5.2 can be applied to eqn (5.12). Hence, system (5.12) is exactly controllable.

5.5 Conclusion

This chapter presents the use of Hilfer fractional derivative to establish the sufficient conditions of exact controllability for Sobolev-type differential equations. This leads to a wider range of stationary states and an additional degree of freedom on the initial condition. The controllability results are carried out using the theory of propagation family, fractional calculus, MNC, and Sadovskii's fixed point approach. In future work, one can apply the obtained results to the various dynamical systems via Hilfer fractional derivative in infinite dimensional spaces.

References

[1] Agarwal, S. Bahuguna, D., "Existence of solutions to Sobolev-type partial neutral differential equations", *J. Appl. Math. Stoch. Anal.*, Art. ID 16308, pp. 10, 2006.

[2] Balachandran, K., Dauer, J. P., "Controllability of functional differential systems of Sobolev-type in Banach spaces", *Kybernetika*, vol. 34, pp. 349–357, 1998.

[3] Balachandran, K., Kiruthika, S., Trujillo, J. J., "On fractional impulsive equations of Sobolev-type with nonlocal condition in Banach spaces",

Computers & Mathematics with Applications, vol. 62, pp. 1157–1165, 2011.

[4] Balachandran, K., Kiruthika S., "Existence of solutions of abstract fractional integrodifferential equations of Sobolev-type", *Computers & Mathematics with Applications*, vol. 64, pp. 3406–3413, 2012.

[5] Barenblat, G., Zheltor, J. Kochiva, I., " Basic concepts in the theory of seepage of homogeneous liquids in fissured rocks", *J. Optim. Theory Appl.*, vol. 24, pp. 1286–1303, 1960.

[6] Deimling, K., *Nonlinear Functional Analysis*, Springer-Verlag, Berlin, 1985.

[7] Fečkan, M., Wang J., Zhou, Y., " Controllability of fractional functional evolution equations of Sobolev-type via characteristic solution operators", *J. Optim. Theory Appl.*, vol. 156, pp. 79–95, 2013.

[8] Furati, K. M., Kassim M. D., Tatar, N. E., "Existence and uniqueness for a problem involving Hilfer fractional derivative", *Computers & Mathematics with Applications*, vol. 64, pp. 1616–26, 2012.

[9] Gou, H., Li, B., " Study on Sobolev-type Hilfer fractional integro-differential equations with delay", *J. Fixed Point Theory Appl.*, vol. 44, pp 1–26, 2018.

[10] Gu, H., Trujillo, J. J., "Existence of mild solution for evolution equation with Hilfer fractional derivative", *Appl. Math. Comput.*, vol. 257, pp. 344–354, 2015.

[11] Heinz, H., *On the behaviour of measures of noncompactness with respect to differentiation and integration of vector valued functions*, Nonlinear Analysis, 7 (1983), vol. 7, 1351–1371, 1983.

[12] Srivastava, H. M., Nain A., K., R. K., , and Das P., "A theoretical study of the fractional-order p-Laplacian nonlinear Hadamard type turbulent flow models having the Ulam–Hyers stability", *Revista de la Real Academia de Ciencias Exactas, Físicas y Naturales. Serie A. Matemáticas*, vol. 17, pp. 160, 2023.

[13] Hilfer, R., *Applications of fractional calculus in Physics*, World Scientific, Singapore, 2000.

[14] Jeet, K., Bahuguna, D., "Approximate Controllability of Nonlocal Neutral Fractional Integro-Differential Equations with Finite Delay", *J. Dyn. Control Syst.*, vol. 22, pp. 485–504, 2016.

[15] Jeet, K., A. Kumar & R. K. Vats. 2023. Approximate controllability of neutral Hilfer fractional differential equations of Sobolev-type in a Hilbert Space. *Math. Control Relat. Fields*, DOI: 10.3934/mcrf.202 3013

[16] Kilbas, A. A., Srivastava, H. M., Trujillo, J. J. *Theory and applications of fractional differential equations*, Amsterdam: Elsevier, (2006).

[17] Kumar, A., Kumar, A., Vats, R. K., "Approximate Controllability of Neutral Delay Integro-Differential Inclusions of Order $\alpha \in (1,2)$ with Non-instantaneous Impulses", *Evo. Equ. Cont. Theory.*, vol. 11, pp. 1635–1654, 2022.

[18] Kumar, A., Jeet, K., Vats, R. K., "Controllability of Hilfer fractional integro-differential equations of Sobolev-type with a nonlocal condition in a Banach space", *Evo. Equ. Cont. Theory.*, vol. 11, pp. 611–627, 2022.

[19] Kumar, A., Vats, R. K., Dhawan K., Kumar, A., "Approximate Controllability of Delay Non-autonomous Integro-Differential System with Impulses", *Mathe. Methods in the Appl. Sci.*, vol. 45 pp. 7322–7335, 2022.

[20] Kumar, A., Vats, R. K., Kumar, A., "Approximate Controllability of second-order non-autonomous system with finite delay", *J. Dyn. Control Syst.*, vol. 26 pp. 611–627, 2020.

[21] Kumar, A., Vats, R. K., Kumar, A., Chalishajar D. N., "Numerical approach to the controllability of fractional order impulsive differential equations". *Demonstratio Mathematica*, vol. 53, pp. 193–207, 2020.

[22] Kumar, P., Vats, R. K., Kumar, A., Kumar, A., "Trajectory Controllability of Integro-Differential Systems of Fractional Order $\gamma \in (1,2]$ in a Banach Space with Deviated Argument". *Springer Proceedings in Mathematics & Statistics*, https://doi.org/10.1007/978-981-16-8177-6_7.

[23] Li, Y., "Existence of solutions to initial value problems for abstract semilinear evolution equations", *Acta Math. Sinica*, vol. 48, pp. 1089–1094 2005.

[24] Liang, J., Xiao, T., "Abstract degenerate Cauchy problems in locally convex spaces", *J. Math. Anal. Appl.*, vol. 259, pp. 398–412, 2001.

[25] Lightbourne, J. H., Rankin, S. M., "M., A partial functional differential equation of Sobolev type", *J. Math. Anal. Appl.*, vol. 93, pp. 328–337, 1983.

[26] Mahmudov, N. I., "Existence and approximate controllability of Sobolev-type fractional stochastic evolution equations", *Bulletin of the Polish Academy of Sciences Technical Sciences*, vol. 62, pp. 205–215, 2014.

[27] Malik M., Kumar A., Controllability of fractional differential equation of order $\varepsilon \in (1, 2]$ with non-instantaneous impulses, *Asian J. of Control*, **113**, (2019), 103–118.

[28] Podlubny, I., *An introduction to fractional derivatives, fractional differential equations, to methods of their solution and some of their applications*, San Diego: Academic Press (2006).

[29] Sharma, O. P. K., Vats, R. K., & Kumar, A. (2023). "A note on existence and exact controllability of fractional stochastic system with finite delay". *International Journal of Dynamics and Control*, vol. 12, pp. 180-189, 2024.

[30] Sharma O P K, Vats R K, Kumar A. New exploration on approximate controllability of fractional neutral-type delay stochastic differential inclusions with non-instantaneous impulse. *Math. Methods Appl. Sci.*, vol. 47, pp. 5161-5190, 2024.

[31] Sharma O P K, Vats R K, Kumar A. Existence and exact controllability results of nonlocal integro-differential neutral stochastic system with finite delay. *The Journal of Analysis*, vol. 32 (1), pp. 573-595, 2024.

[32] Wang, J., Fečkan, M., Zhou, Y., "Controllability of Sobolev type fractional evolution systems", *Dynamics of PDE,* , vol. 11, pp. 71–87, 2014.

[33] Yadav,V., R. K. Vats, A. Kumar, and K. Jeet, 2023 New exploration on the existence and null controllability of fractional Hilfer stochastic systems driven by Poisson jumps and fractional Brownian motion with non-instantaneous impulse. *International Journal of Dynamics and Control*, https://doi.org/10.1007/s40435-024-01451-2.

[34] Yadav,V., R. K. Vats, A. Kumar, and K. Jeet, 2023 Results on the existence and approximate controllability of neutral-type delay integro-differential system with noninstantaneous impulse. *Math. Methods Appl. Sci.*, vol. 46, pp. 16992–17009, 2023.

6

Mathematical Modeling using Fuzzy Petri Nets and its Classifications

Riddhi Jangid and Gajendra Pratap Singh*

School of Computational and Integrative Sciences, Jawaharlal Nehru University, India
E-mail: riddhi40_sit@jnu.ac.in; gajendra@jnu.ac.in
*Corresponding author

Abstract

Petri nets have been evolving since the 1960s when they were introduced in discrete mathematics. They are a graph theoretical tool used to model discrete event dynamic systems. Many extensions of Petri nets have been proposed depending on the system requirements and applicability in the engineering domain. The fuzzy Petri net (an extension of Petri nets) was first introduced in 1988. It combines fuzzy logic and reasoning to study knowledge-based systems where there is some uncertainty. Since theoretical research lags in the field, this chapter explains the basics of fuzzy Petri nets, and their classifications with an introductory application.

Keywords: Fuzzy logic, fuzzy production rules, Petri nets, knowledge-based systems

6.1 Introduction

Zadeh introduced the fuzzy set theory concept in 1965 [Zadeh (2023)]. It is an extension of the classical crisp set theory that introduces the concept of uncertainty in sets. It extends the decision boundaries that reflect more accuracy for modeling in any field of application. Then we express such models by some defined *fuzzy rules*. The fuzzy rules represent the relationships between inputs and outputs of any process in knowledge-based systems or we say systems with uncertainty. Hence motivated to look into the modeling from fuzzy rules, we study an emerging graph theoretical modeling tool, Petri net (PN). Petri nets (PNs), introduced by Carl Adam Petri in his PhD thesis during the 1960s, are specifically used to learn the behavior of discrete events and dynamical systems' is describing the type of systems we are describing here in the context. The applications of Petri nets cover a large number of such systems from different domains including biological systems, manufacturing systems, communication networks, etc. [Singh & Gupta (2019), Singh et al. (2020), Singh & Jha (2022), Jangid et al. (2023)]. Depending upon the specifications and requirements of the systems, many functions were associated with the Petri net graph and they are called high-level Petri nets (HLPNs). HLPNs contain time Petri nets, colored Petri nets, and stochastic Petri nets. A fuzzy Petri net (FPN) is one such type that represents the fuzzy information in their PN modeling. FPNs were first introduced by Lipp in 1984 for controlling complex industrial processes. Later on, high-level FPNs (HLFPNs) were introduced combining multiple characteristics into a model and different classifications of FPNs were formed [Jiang et al. (2022)].

In the chapter, we have introduced the existing literature on FPNs from a different perspective and an example of its application that have not been reported so far. The second section describes the topic's preliminaries and fundamentals and then the third section discusses fuzzy Petri nets' classifications. The fourth section explains examples of the applications to the topic and the last section concludes with the discussion and future directions.

6.2 Preliminaries

A fuzzy set is an extension of a crisp set whose every element has a degree of membership. If we denote a fuzzy set by $(A,\ \mu_A)$, then A is a set and μ_A is a real value that maps from A to $[0, 1]$. Further, $\mu_A(x)$ is called the grade of membership x, $\forall\ x\ \in\ A$. Definition 6.1 formalizes the concept mathematically.

Definition 6.1. (Fuzzy set) A fuzzy set is denoted as

$$A = \mu_A(x_1)/x_1 + \cdots + \mu_A(x_n)/x_n,$$

where $A\ \subseteq\ \{x_1,\ x_2,\ \ldots\ ,\ x_n\}\ =\ X$ and $\mu_A(x_i)/x_i$ is a pair "grade of membership element."

While modeling, the first thing to consider is the system rules and requirements for the Petri net structure. Concerning this, we first discuss *fuzzy rules* [Liu et al. (2017)] that are used to capture the fuzziness in decision systems. Usually, every rule is expressed in fuzzy $IF - THEN$ rule form where both input and output values are fuzzy values denoted by fuzzy sets. A fuzzy knowledge-based system can be thought of as a computer program that works on the idea of AI. It performs approximate reasoning using vague facts and vague rules. If X_1 is a fact in the fuzzy knowledge base, then the rule in the fuzzy knowledge-based system says: $IF\ X_1\ THEN\ Y$ (CF), where CF means certainty factor.

An illustrative example of the mentioned concepts is discussed here. Let us suppose we have 10 employees in a company with different heights and weights, i.e., $X = \{x_1, x_2, x_3, x_4, x_5, x_6, x_7, x_8, x_9, x_{10}\}$. The question is, who is "tall" relative to the others? This defines the different membership values of all the employees. For example, the tallest employee gets $\mu_A(x_1) = 1$ while the others get their relative membership values less than 1 accordingly. And considering an ideal condition where everyone has the same BMI, then the rule says IF height is tall, $THEN$ weight is more.

We further use these rules while modeling an FPN. FPN is an extension of the concept of Petri net. PN [Murata (1989), Peterson (1981)] is a modeling tool used for modeling and analyzing communications systems, manufacturing processes, etc. A graphical model of a process is prepared using *places* (generally represented using a circle), *transitions* (generally represented using a rectangle/vertical line), and *arcs* (generally represented using a directed arrow) being the three elements such that places and transitions are connected using arcs and no place (transition) is connected to another place (transition) [5]. These places hold a token inside them which is depicted using a *token* and represented using a dot inside the place. The process is then analyzed using these parameters and its behavior is studied. The same concept we use to define FPNs but with uncertainty in the systems. The fundamental definition of the classical fuzzy Petri net (FPN) that has been adopted by many of the researchers is illustrated in Definition 6.2.

Definition 6.2. (Fuzzy Petri net) FPN is a 9-tuple: FPN $= (P, T, D, I, O, f, \alpha, \beta)$, where

$P = \{P_1, P_2, P_3, ..., P_m\}$ represents a finite set of m places,
$T = \{T_1, T_2, T_3, ..., T_n\}$ represents a finite set of n transitions,
$D = \{D_1, D_2, D_3, ..., D_m\}$ represents a finite set of m propositions,
$I : T \rightarrow P^{\infty}$ represents an input function mapping from the set of transitions to the bag of places,

$O : T \rightarrow P^{\infty}$ represents an output function mapping from the set of transitions to the bag of places,
$f : T \rightarrow [0, 1]$ represents an association function from the set of transitions to real numbers,
$\alpha : P \rightarrow [0, 1]$ represents an association function from the set of places to real numbers, and
$\beta : P \rightarrow D$ represents an association function that is a bijective mapping from the set of places to the set of propositions.

P, T, and D are three disjoint sets where $|P| = |D|$, I and O represents the set of directed arcs joining P and T. If $f(t_i) = \mu_i$, it represents transition t_i , which is associated with a threshold value of μ_i belonging between 0 and 1. If $\beta\ (p_i) = d_i$, then the proposition d_i holds for the place p_i. The marking of the net is denoted by $\alpha\ (p_i)$ for every place p_i. It is important here to note that the marking of any place lies between the real value 0 and 1 only. It indicates the degree of truth of the respective proposition d_i. For instance, if $\alpha\ (p_i) = x_i$ and $\beta\ (p_i) = d_i$, then the degree of truth of d_i proposition is x_i [Gupta et al. (2019)]. An example of the fuzzy production rule that can be modeled using FPN in the knowledge representation system is shown in Figure 6.1. It explains the $Rule :\ IF\ X_1\ THEN\ Y$ (CF = μ_1).

Consider an example where we have a water valve established. It has two pipes incoming, one for hot water and the other for cold water. The function of the water valve is to regulate the temperature and water flow in the pipes. Now, for simplicity, let us suppose the temperature of the water outflow can be high, medium, or low. Also, three states of the water valve can be either fully hot, fully cold, or half hot half cold. It is important to note here that since the classifications are not precise, we cannot say about the exact state. Now we need to model the situation to understand the possibility of all the outcomes.

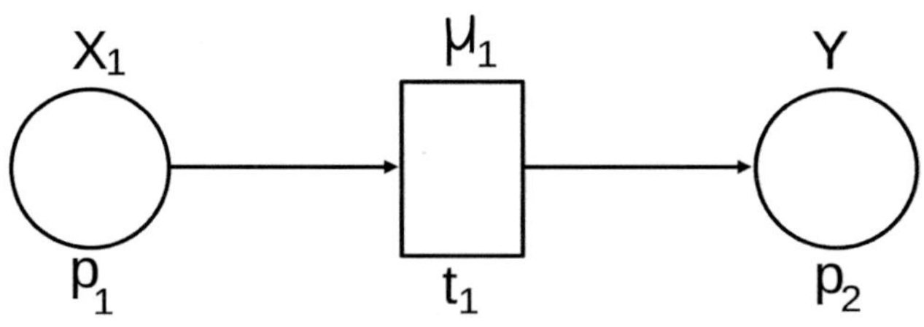

Figure 6.1 Fuzzy IF–THEN rule modeled using fuzzy Petri net.

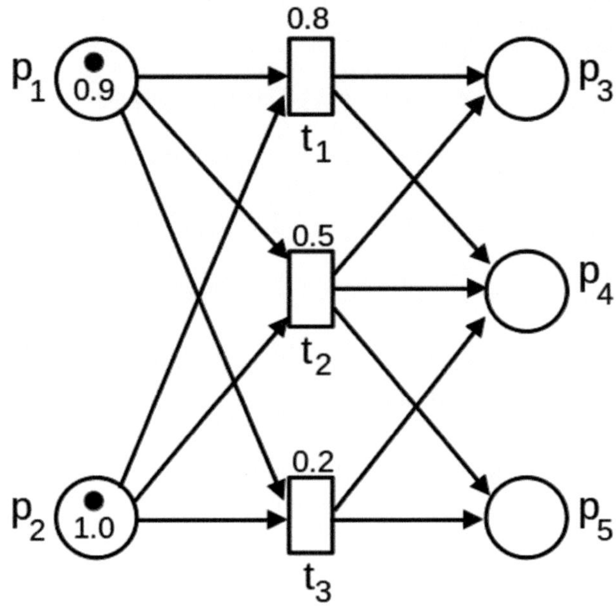

Figure 6.2 Example of FPN modeling.

The parameters of the FPN shown in Figure 6.2 are defined as follows.

$$P = \{p_1,\ p_2,\ p_3,\ p_4,\ p_5\},$$
$$T = \{t_1,\ t_2,\ t_3\},$$
$$D = \{d_1,\ d_2,\ d_3, d_4,\ d_5\},$$
$$I(t_1) = \{p_1,\ p_2\}, I(t_2) = \{p_1,\ p_2\},\ I(t_3) = \{p_1,\ p_2\}$$
$$O(t_1) = \{p_3,\ p_4\},\ O(t_2) = \{p_3,\ p_4,\ p_5\},\ O(t_3) = \{p_4,\ p_5\}$$
$$f(t_1) = 0.8, f(t_2) = 0.3, f(t_3) = 0.2$$
$$\alpha(p_1) = 0.9, \alpha(p_2) = 1.$$

Table 6.1 explains the representation of all the parameters we have considered in the example.

To solve the problem, the firing rule is used. The firing rule in FPN is, a transition t_i is said to be enabled if $\forall\ p_j\ \in\ I(t_i)$, $\alpha(p_j) \geq \lambda$ where λ is a threshold value that lies in $[0, 1]$. In the above example, if the transitions are enabled and ready to fire, the calculation of the new marking in every output place of transition t_i can be calculated using $\alpha(p_k) = \min(x_1,\ x_2,\ ...,\ x_i) *$

Table 6.1 Description of the FPN example parameters.

Parameter of FPN	**Representation**
$\beta(p_1) = d_1$	Water coming is hot from the hot pipe
$\beta(p_2) = d_2$	Water coming is cold from the cold pipe
$\beta(p_3) = d_3$	High water temperature of the outflow
$\beta(p_4) = d_4$	Medium water temperature of the outflow
$\beta(p_5) = d_5$	Low water temperature of the outflow
t_1	The water valve is in a *fully hot* state
t_2	The water valve is in *a half-hot, half-cold* state
t_3	The water valve is in a *fully cold* state

$\mu_i = x_k$ where $f(t_i) = \mu_i$ and $\alpha(p_j) = x_i, \forall\, p_k \in O(t_i)$ and $p_j \in I(t_i)$. Other firing rules can be referred to from [Manoj et al. (1998)].

In the above example, if $\lambda = 0.5$, then transitions t_1 and t_2 are enabled and firing of transition t_1 gives us the marking $\alpha(p_3) = 0.9 * 0.8 = 0.72$, $\alpha(p_4) = 0.9 * 0.5 = 0.45$ and the remaining markings are 0 since only p_3 and p_4 are the output places of the transition t_1. The marking shows us the water temperature of the outflow is high with possibility degree of 0.72 and medium with possibility degree of 0.45.

6.3 Classifications of Fuzzy Petri Net

More generalizations were added in the Petri nets that improved the modeling power of the real systems. HLPNs were merged with the concept of FPNs and they were further named as HLFPNs. We are discussing a few classifications here.

6.3.1 Fuzzy time Petri net

Introducing time variables in the form of interval or duration with Petri nets gives us a variety of possibilities that can be applied to model systems by considering time as a parameter. The time variable introduced here is deterministic. Merlin [Popova-Zeugmann (2013)] described time Petri nets by introducing time intervals $[a_t, b_t]$ attached to the transition t such that no transition can fire before a_t time units and after b_t time units. We call a_t and b_t the earliest firing time of transition *t* and the latest firing time of transition t respectively. Whereas, Ramchandani introduced timed Petri nets by assigning a time duration s_t to the transition t. We consider the latter one and define the time fuzzy Petri nets as discussed in Definition 6.3.

Definition 6.3. (Fuzzy time Petri net) FTPN is 11-tuple: FTPN $=$ $(P, T, D, I, O, f, \alpha, \beta, S,\ TS, M_0)$, where

$P, T, D, I, O, f, \alpha, \beta$ represent the same as defined in Definition 6.2.

$S = \{s_1, s_2\ , \ldots, s_n\},\ S : T\ \rightarrow R^+$ represents a time delay function that is associated with every transition.

TS represents a set of transition states that equals 1 if any transition has fired and 0 if it has not.

M_0 represents the initial marking of PN.

6.3.2 Fuzzy stochastic Petri net

Stochastic Petri nets (SPNs) are a part of time-dependent Petri nets where the firing times associated with the transitions are exponentially distributed. In other words, a random delay in firing is associated with every transition that is negatively exponential in most of the cases. The basic idea behind introducing the type of Petri net is to obtain a connection between a Petri net and continuous time Markov chains. The fuzzy concept used with the stochastic Petri nets is defined as *fuzzy stochastic Petri net* as discussed in Definition 6.4.

Definition 6.4. (Fuzzy stochastic Petri net) FSPN is 6-tuple: FSPN $=$ $(P,\ T,\ I,\ O,\ M_0,\ \Lambda)$ where $P,\ T,\ I,\ O,\ M_0$ are the same as in Definition 6.3. And Λ represents a function that associates to every transition a real value, denoting a fuzzy set of transition firing rates.

6.3.3 Fuzzy colored Petri net

The tokens in colored Petri net (CPN) are assigned with some data values that they carry with themselves and are called token color. Given a place, all tokens in that place have token colors that belong to a specific type, which we call the color set of that place. Different places can also have the same color sets. Marking in CPNs is the tokens that are distributed in places [Jensen (2013)]. The underlined markings besides the places are the initial markings in these Petri nets. Practically, we cannot just put tokens inside the places to show the current marking of colored Petri nets as we did in previous Petri nets because we also need to show the token colors in this case. As a result of introducing token colors and color sets, a lesser number of places are used in comparison to normal Petri nets. Now, the enabling of transitions will depend on the type of token colors that are in the input places of the transitions, which will affect the output tokens as well. To remove the complexity, arc expressions were introduced. These arc expressions specify the collection of the tokens and contain several variables that are associated with the transitions. To combine many different processes in a system, along with the places one needs to find a way to combine transitions as well, i.e., a transition can perform different actions. The variables introduced with the arc expressions will solve our purpose by evaluating different values. We look at the formal definition of Fuzzy colored Petri net where the fuzziness is introduced along with the concept of colored nets.

Definition 6.5. (Fuzzy colored Petri net)[1] FCPN is a 10-tuple: FCPN $= (P, T, I, O, \Sigma, c, g, f, v, m_0)$ where P, T, I, O are the same as in Definition 6.2. And

Σ: represents a finite set of colour sets,
$c : P \rightarrow \Sigma$ represents a color function where $\forall\, p \in P$ is a color set $c(p) \in \Sigma$,
$g : T \rightarrow EXP$ represents a guard function where $\forall\, t \in T$, a Boolean-type guard expression is assigned,
$f : A \rightarrow EXP$ represents an arc function where $\forall\, a \in A$ is an arc expression of a multiset type $c(p)_{MS}$ (p is associated with a),
$m_0 : P \rightarrow EXP$ represents initialization assigned to every place $p \in P$, an initialization expression of a multiset type $c(p)_{MS}$, and
$v : T \rightarrow H_c$ represents the assigned firing rate for each transition.

6.3.4 Fuzzy continuous Petri nets (FCPNs)

Unlike a classical PN, continuous Petri nets (CPNs) are those where the marking is a positive real number instead of a natural number. One can think of discrete tokens converting into continuous tokens in which the tokens are splitting into equal parts, say k, and this k is tending toward infinity. This $1/k$ part is what we call a token in this case. But the reachability graph obtained here by firing will be infinite. Hence, we consider initial marking particularly called "mark" instead of tokens [3]. Also, transitions here fire continuously. Hence, we denote firing of transition times simultaneously (means in one go) by $[t_i]^{\kappa}$, where κ is a non-negative and non-integral value.

Definition 6.6. (Fuzzy continuous Petri net) FCPN is 7-tuple: FCPN $= (P, T, I, O, f, v, m_0)$ where

P represents a finite set of continuous places,
T represents a finite set of continuous transitions,
I, O, and v are same as defined in Definition 6.5,

f represents a function mapping from the set of arcs to a non-negative real number, and

m_0 represents the initial marking assigned to the places.

Besides these, there are many other classifications like generalized fuzzy Petri nets (GFPNs), fuzzy time stochastic Petri nets, fuzzy colored stochastic Petri nets, intuitionistic fuzzy Petri nets, etc. But the idea behind defining them follows the same formalism as we discussed. For example, GFPNs include the arc weights and belief strengths while modeling. These models help in analyzing real-life problems that we will discuss in the next section.

6.4 Applications

FPNs and their classifications have been applied in many fields depending on the system requirements. The list of classifications increases with different modeling ideas. The most commonly applied areas are as follows. Yang et al. (2018) discussed the applications in industrial processes using fuzzy timed Petri nets, Liu et al. (2021) studied biological systems by developing an FCPN model for its analysis, 2 studied flexible manufacturing systems, Zhou and Zain (2016) discussed the turbine fault diagnosis system for fault detection, Hajduk and Wojtowicz (2016) studied traffic control systems by proposing a different FPN formalism, Zhou et al. (2017) proposed a weighted FPN model for risk assessment in chemical industries, etc.

6.5 Conclusion and Future Scope

Introduction of the fuzzy logic in Petri nets gave rise to a new field in mathematics that is still emerging and has a lot more scope to improve. FPNs have been mainly used in knowledge representation and reasoning algorithms and some industrial applications, but they lack parallel reasoning and their learning algorithms. They have also been looked at from the perspective of State Space Explosion problems recently. Analyzing these nets by graphical methods and algebraic methods has been explored in the past few decades. We have discussed here the very fundamental concepts that help readers to get a quick introduction to the topic and they can apply the concepts to model new applications where there is some vagueness in the systems. One new example is also discussed in the chapter here. Readers are encouraged to study the reachability graph that we generate with the FPNs and interpret them also. In the big data era, the FPN representation is also not applicable in the management field. Exploring in that direction is also encouraged.

Acknowledgment

Gajendra Pratap Singh acknowledges the support from the Department of Science and Technology (DST)-Science and Engineering Research Board (SERB) project (MATRICS ID: **MTR/2021/000378**), M.Sc. consumable grant, DBT BIC and the Department of Biotechnology, Ministry of Science & Technology, Govt. of India (Project ID: BT/PR40251/BITS/137/11/2021), and ICMR project ID 2021-15450 to provide the research facility. The authors would also like to thank the reviewers and readers for their valuable feedback.

References

[1] Assaf, G., Heiner, M., & Liu, F. (2022). Coloured fuzzy Petri nets for modelling and analysing membrane systems. *Biosystems*, *212*, 104592.

[2] Chen, Z., Xu, B., Lu, N., & Song, W. (2019, November). A fault diagnosis method for flexible manufacturing tool system based on fuzzy Petri net. In *2019 International Conference on Intelligent Informatics and Biomedical Sciences (ICIIBMS)* (pp. 374-379). IEEE.

[3] David, René, and Hassane Alla. Discrete, continuous, and hybrid Petri nets. Vol. 1. Berlin: Springer, 2005.

[4] Gupta, S., Kumawat, S., & Singh, G. P. (2019). Fuzzy Petri net representation of fuzzy production propositions of a rule based system. In Advances in Computing and Data Sciences: Third International Conference, ICACDS 2019, Ghaziabad, India, April 12–13, 2019, Revised Selected Papers, Part I 3 (pp. 197-210). Springer Singapore.

[5] Jangid, R., & Singh, G. P. (2023). Construction of m-cube mesh networks using Petri nets. *International Journal of System Assurance Engineering and Management*, *14*(Suppl 1), 256-262.

[6] Jangid, R., Somvanshi, P., & Singh, G. P. (2023). Graph Theory in the Biological Networks. In *Biological Networks in Human Health and Disease* (pp. 1-14). Singapore: Springer Nature Singapore.

[7] Jensen, Kurt. Coloured Petri nets: basic concepts, analysis methods and practical use. Vol. 1. Springer Science & Business Media, 2013.

[8] Jiang, W., Zhou, K. Q., Sarkheyli-Hägele, A., & Zain, A. M. (2022). Modeling, reasoning, and application of fuzzy Petri net model: a survey. *Artificial Intelligence Review*, *55*(8), 6567-6605.

[9] Liu, F., Sun, W., Heiner, M., & Gilbert, D. (2021). Hybrid modelling of biological systems using fuzzy continuous Petri nets. *Briefings in Bioinformatics*, *22*(1), 438-450.

[10] Liu, H. C., You, J. X., Li, Z., & Tian, G. (2017). Fuzzy Petri nets for knowledge representation and reasoning: A literature review. *Engineering Applications of Artificial Intelligence*, *60*, 45-56.

[11] Manoj, T. V., Leena, J., & Soney, R. B. (1998). Knowledge representation using fuzzy Petri nets-revisited. *IEEE transactions on knowledge and data engineering*, *10*(4), 666-667.

[12] Murata, T. (1989). Petri nets: Properties, analysis and applications. *Proceedings of the IEEE*, *77*(4), 541-580.

[13] Peterson, J. L. (1981). *Petri net theory and the modeling of systems.* Prentice Hall PTR.

[14] Popova-Zeugmann, L., & Popova-Zeugmann, L. (2013). *Time petri nets* (pp. 31-137). Springer Berlin Heidelberg.

[15] Singh, G. P., & Gupta, A. (2019, March). A Petri net analysis to study the effects of diabetes on cardiovascular diseases. In *2019 6th International Conference on Computing for Sustainable Global Development (INDIACom)* (pp. 481-488). IEEE.

[16] Singh, G. P., & Jha, M. (2022). Modeling the Immune Response of B-Cell Receptor Using Petri Net for Tuberculosis. In *Advanced Computational Techniques for Sustainable Computing* (pp. 259-272). Chapman and Hall/CRC.

[17] Singh, G. P., Jha, M., Singh, M., & Naina. (2020). Modeling the mechanism pathways of first line drug in Tuberculosis using Petri nets. *International Journal of System Assurance Engineering and Management*, *11*, 313-324.

[18] Yang, B., & Li, H. (2018). A novel dynamic timed fuzzy Petri nets modeling method with applications to industrial processes. *Expert Systems with Applications*, *97*, 276-289.

[19] Zadeh, L. A. (2023). Fuzzy logic. In *Granular, Fuzzy, and Soft Computing* (pp. 19-49). New York, NY: Springer US.

[20] Zhou, J., Reniers, G., & Zhang, L. (2017). A weighted fuzzy Petri-net based approach for security risk assessment in the chemical industry. *Chemical Engineering Science*, *174*, 136-145.

[21] Zhou, K. Q., & Zain, A. M. (2016). Fuzzy Petri nets and industrial applications: a review. *Artificial Intelligence Review*, *45*, 405-446.

7

Exploring New Exact Solutions of (2 + 1)-Dimensional Date–Jimbo–Kashiwara–Miwa Equation with $\tan\left(\frac{\phi(\xi)}{2}\right)$-Expansion Method

Lakhveer Kaur and Pallavi Verma

Department of Mathematics, Jaypee Institute of Information Technology, India
E-mail: lakhveer712@gmail.com; pallavi4verma@gmail.com

Abstract

The current course of work involves analytical study of the $(2+1)$-dimensional Date–Jimbo–Kashiwara–Miwa equation. One of the effective techniques, known as the $\tan\left(\frac{\phi(\xi)}{2}\right)$-expansion method, has been executed uniformly in order to attain new exact solutions. Numerous kinds of exact solutions consisting of arbitrary constants are established in terms of hyperbolic, exponential, and trigonometric functions. The results affirm that such expansion methods are the best tool for handling a wide variety of high-dimensional nonlinear wave equations.

Keywords: Date–Jimbo–Kashiwara–Miwa equation, $\tan\left(\frac{\phi(\xi)}{2}\right)$-expansion method, exact solutions, hyperbolic function, exponential function, trigonometric function

7.1 Introduction

The diverse range of discipline essentially involves analyzing and forecasting the changes that take place all across the universe, leading to the development of some of the most significant concepts, and applications for each and every arena of science and engineering in terms of nonlinear partial differential equations (NLPDEs). Solutions to NLPDEs are frequently used in various courses as a potential tool for recognizing the core characteristics of the concerned phenomenon. Analytic solutions enable researchers to execute intense studies to identify the parameters with a setting of adequate natural conditions. The grapple to find these analytical, exact explicit and implicit solutions is fraught with difficulties, such as trouble recognizing the solution, singularities, and convergence considerations that lead to no solution. However, scientists and academicians are very well successful to discover an analytical solution to many NLPDEs [1–5].

Many scientists and researchers have worked tirelessly in recent years to investigate and produce analytic answers to NLPDEs, which have been designated as one of the most important parts of science and most intriguing scientific topics. For comprehending the qualitative properties of multiple occurrences and processes in a variety of natural science disciplines, mathematical optical soliton solutions are essential. Analytic solutions are also widely used in many courses as real-world illustrations that highlight key ideas in a theory and facilitate mathematical explanation. A graphical representation of solutions for NLPDEs helps to illustrate the variety of sophisticated nonlinear phenomena, such as: the spatial localization of transfer processes, the propagation of waves, etc. Therefore, owing to the massive significance of optical solitons and other solutions, we get inspired to seek a unique class of solitons for the (2+1)-dimensional Date–Jimbo–Kashiwara–Miwa equation using $\tan\left(\frac{\phi(\xi)}{2}\right)$ in this study. The method has been employed to do the job in the present study due to its proven effectiveness and efficiency in establishing various types of solutions to the nonlinear differential equations.

An integral prolongation of the Kadomstev–Petviashvili hierarchy, namely the $(2+1)$-dimensional Date–Jimbo–Kashiwara–Miwa equation [6–10] is read as

$$v_{xxxxy} + 4v_{xxy}v_x + 2v_{xxx}v_y + 6v_{xy}v_{xx} - \epsilon_1 v_{yyy} - 2\epsilon_2 v_{xxt} = 0, \quad (7.1)$$

where $v = v(x, y, t)$.

Wang et al. [11] discussed integrability properties and multi-shock wave solutions for of the $(2+1)$-dimensional Date–Jimbo–Kashiwara–Miwa equation in a elaborative manner, both mathematically and graphically. Also, generalized bilinear system of the considered equation has been derived along with the Lax pair and infinite conservation laws [11]. Cheng et al.

[12] guaranteed the localization as well as analyticity of the lump solutions for the considered equation via a non-zero determinant condition. Ismael et al. [13] explored various kinds of solutions, namely multi-soliton, multi-fusion, and multi-lump solutions with the help of Hirota bilinear method and long-wave method. Furthermore, the phenomena of physical collision of different kinds of the obtained solutions were also depicted [13]. Classical Lie symmetry method has been applied on the $(2+1)$-dimensional Date–Jimbo–Kashiwara–Miwa equation in order to obtain symmetries and possible independent vector fields by virtue of invariance constraint of the group-theoretic techniques [14]. The inhomogeneities of media motivated Wazwaz to extend eqn (7.1) to a variable coefficient one [15].

The arrangement of this chapter has been done as follows. Introduction and literature review connected with the $(2+1)$-dimensional Date–Jimbo–Kashiwara–Miwa equation is cited in Section 7.1. The step-wise detailing of the $\tan\left(\frac{\phi(\xi)}{2}\right)$-expansion method has been given in Section 7.2. Application of the $\tan\left(\frac{\phi(\xi)}{2}\right)$-expansion method on the considered equation leading to new exact solutions is mentioned in the section "Exact solutions" is mentioned in the Section 7.3. Finally, some concluding remarks have been mentioned in Section 7.4.

7.2 Methodology

This section of the chapter is dedicated to a gradual description of $\tan\left(\frac{\phi(\xi)}{2}\right)$-expansion method [16–20], for possible implementation to the $(2+1)$-dimensional Date–Jimbo–Kashiwara–Miwa equation, so that its exact solutions can be procured:

Step 1: The following nonlinear partial differential equation

$$K\left(w, \frac{\partial w}{\partial x}, \frac{\partial w}{\partial t}, \frac{\partial^2 w}{\partial x^2}, \frac{\partial^2 w}{\partial t^2}, w\frac{\partial w}{\partial x}, ...\right) = 0 \tag{7.2}$$

can be remodeled as an ordinary differential equation (ODE):

$$L(w, w', -\alpha w', w'', \alpha^2 w'', ww', -\alpha ww', ...) = 0, \tag{7.3}$$

by recasting of independent and dependent variables as: $w(x,t) = w(\xi)$, $\xi = x - \lambda t$. Here, λ is a free parameter, which will be evaluated concurrently in the mean process.

Step 2: The ODE (7.3) is assumed to have solution of following type

$$w(\xi)=\sum_{i=0}^{m}G_i\left(k+\tan\left(\frac{\phi(\xi)}{2}\right)\right)^i+\sum_{i=1}^{m}H_i\left(k+\tan\left(\frac{\phi(\xi)}{2}\right)\right)^{-i}, \tag{7.4}$$

where $G_i(i=0,1,\ldots,m)$, $H_i(i=1,2,\ldots,m)$ are the parameters that will be explored in a way that G_m and H_m are not zero simultaneously and the function $\phi=\phi(\xi)$ is defined as a solution of the subsequent ODE:

$$\frac{d\phi(\xi)}{d\xi}=p_1\sin(\phi(\xi))+p_2\cos(\phi(\xi))+p_3. \tag{7.5}$$

The particular categories, obtained as solutions of ODE (7.5), are stated below:

Family 1: $p_2-p_3\neq 0$ and $p_1^2+p_2^2-p_3^2<0$,
implies $\phi(\xi)=2\arctan\left[\frac{p_1}{p_2-p_3}-\frac{\sqrt{p_3^2-p_2^2-p_1^2}}{p_2-p_3}\tan\left(\frac{\sqrt{p_3^2-p_2^2-p_1^2}}{2}(\xi+C)\right)\right]$.

Family 2: $p_2-p_3\neq 0$ and $p_1^2+p_2^2-p_3^2>0$,
implies $\phi(\xi)=2\arctan\left[\frac{p_1}{p_2-p_3}+\frac{\sqrt{p_2^2+p_1^2-p_3^2}}{p_2-p_3}\tan\left(\frac{\sqrt{p_2^2+p_1^2-p_3^2}}{2}(\xi+C)\right)\right]$.

Family 3: $p_1^2+p_2^2-p_3>0$, $p_2\neq 0$ and $p_3=0$,
implies $\phi(\xi)=2\arctan\left[\frac{p_1}{p_2}+\frac{\sqrt{p_2^2+p_1^2}}{p_2}\tan\left(\frac{\sqrt{p_2^2+p_1^2}}{2}(\xi+C)\right)\right]$.

Family 4: $p_1^2+p_2^2-p_3<0$, $p_3\neq 0$ and $p_2=0$,
implies $\phi(\xi)=2\arctan\left[\frac{-p_1}{p_3}+\frac{\sqrt{p_3^2-p_1^2}}{p_3}\tan\left(\frac{\sqrt{p_3^2-p_1^2}}{2}(\xi+C)\right)\right]$.

Family 5: $p_1^2+p_2^2-p_3>0$, $p_2-p_3\neq 0$ and $p_1=0$,
implies $\phi(\xi)=2\arctan\left[\sqrt{\frac{p_2+p_3}{p_2-p_3}}\tanh\left(\frac{\sqrt{p_2^2-p_3^2}}{2}(\xi+C)\right)\right]$.

Family 6: $p_1^2+p_2^2=p_3^2$, implies $\phi(\xi)=-2\arctan\left(\frac{(p_2+p_3)(p_1(\xi+C)+2)}{p_1^2(\xi+C)}\right)$.

Family 7: $p_1=p_2=p_3=qp_1$, implies $\phi(\xi)=\arctan\left(\exp\left(qp_1(\xi+C)\right)-1\right)$.

Family 8: $p_1=p_3=qp_1$ and $p_2=-qp_1$, implies $\phi(\xi)=$
$\arctan\left(\frac{\exp(qp_1(\xi+C))}{1-\exp(qp_1(\xi+C))}\right)$.

Family 9: $p_3=p_1$, implies $\phi(\xi)=-2\arctan\left(\frac{(p_1+p_2)\exp(p_2(\xi+C))-1}{(p_1-p_2)\exp(p_2(\xi+C))-1}\right)$.

Family 10: $p_1 = p_3$, implies $\phi(\xi) = 2\ \arctan\left(\frac{(p_2+p_3)\exp(p_2(\xi+C))+1}{(p_2-p_3)\exp(p_2(\xi+C))-1}\right)$.

Family 11: $p_3 = -p_1$, implies $\phi(\xi) = 2\ \arctan\left(\frac{p_2+\exp(p_2(\xi+C))-p_1}{\exp(p_2(\xi+C))-p_2-p_1}\right)$.

Family 12: $p_2 = -p_3$, implies $\phi(\xi) = -2\ \arctan\left(\frac{\exp(p_1(\xi+C))p_1}{p_3\exp(p_1(\xi+C))-1}\right)$.

Family 13: $p_1 = p_3$ and $p_2 = 0$, implies $\phi(\xi) = -2\ \arctan\left(\frac{(\xi+C)p_3+2}{p_3(\xi+C)}\right)$.

Family 14: $p_2 = p_3$ and $p_1 = 0$, implies $\phi(\xi) = 2\ \arctan(p_3(\xi+C))$.

Family 15: $p_2 = -p_3$ and $p_1 = 0$, implies $\phi(\xi) = -2\ \arctan\left(\frac{1}{p_3(\xi+C)}\right)$.

Family 16: $p_2 = p_3$, implies $\phi(\xi) = 2\ \arctan\left(\frac{\exp p_1(\xi+C)-p_3}{p_1}\right)$.

Family 17: $p_1 = 0$ and $p_3 = 0$,
implies $\phi(\xi) = \arctan\left(\frac{e^{2p_2(\xi+C)}-1}{e^{2p_2(\xi+C)}+1}\frac{2e^{p_2(\xi+C)}}{e^{2p_2(\xi+C)}+1}\right)$.

Family 18: $p_2 = 0$ and $p_3 = 0$,
implies $\phi(\xi) = \arctan\left(\frac{2e^{p_1(\xi+C)}}{e^{2p_1(\xi+C)}+1}, -\frac{e^{2p_1(\xi+C)}-1}{e^{2p_1(\xi+C)}+1}\right)$,
where p_1, p_2, p_3, q, and C are parameters. Making homogenous balance among the maximum order nonlinear terms and the maximum order partial derivative terms appearing in eqn (7.3) leads to the value of m.

Step 3: Considering eqn (7.4) with appropriate value of m into eqn (7.3) and thenafter gathering coefficients of all distinct powers of $\tan\left(\frac{\phi(\xi)}{2}\right)$ and $\cot\left(\frac{\phi(\xi)}{2}\right)$, furthermore equating each of that coefficient with zero, we managed to get an over-determined nonlinear algebraic equation for $G_0, G_i, H_i (i = 1, 2, ..., m), p_1, p_2, p_3, \lambda$, and k.

Step 4: Finally, analytic solutions of eqn (7.2) are furnished by reverting back to original variables x, t along with substitution of constant values as in (7.4) and employing particular categories of solutions for ODE (7.5).

7.3 $(2+1)$-Dimensional Date–Jimbo–Kashiwara–Miwa Equation

The $(2+1)$-dimensional Date–Jimbo–Kashiwara–Miwa equation (7.1) is converted to the subsequent ODE:

$$6\alpha_1^3\alpha_2 u'u'' + \alpha_2\alpha_1^4 u'''' + 2\alpha_1^2\alpha_3\epsilon_2 u'' - \alpha_2^3\epsilon_1 u'' = 0, \qquad (7.6)$$

with the execution of variable transformation as given below:

$$v(x,y,t) = u(\xi),\ \xi = \alpha_1 x + \alpha_2 y - \alpha_3 t, \tag{7.7}$$

where $\alpha_1,\ \alpha_2$, and α_3 are free constants. On balancing $u'u''$ with u'''' in eqn (7.6) gives $m = 1$.

Putting the value of $m = 1$ and $k = 0$ in eqn (7.4), the solution for eqn (7.6) is written as

$$u(\xi) = G_1 \tan\left(\frac{\phi(\xi)}{2}\right) + G_0 + H_1 \cot\left(\frac{\phi(\xi)}{2}\right). \tag{7.8}$$

Clubbing eqn (7.8) with eqn (7.6), thenafter the coefficients of all distinct powers of $\cot\left(\frac{\phi(\xi)}{2}\right)$ and $\tan\left(\frac{\phi(\xi)}{2}\right)$ are assembled. A cluster of nonlinear algebraic equations are perceived, by equating different coefficients to zero, which are worked out for involved parameters. Eventually, we are in the position to recover solutions of eqn (7.1) in terms of original variables, as stated in succeeding cases:

Case 1:

$$\begin{aligned} p_1 &= 0,\ G_1 = p_2\alpha_1 - p_3\alpha_1,\ H_1 = p_2\alpha_1 + p_3\alpha_1, \\ \epsilon_1 &= \frac{2\alpha_1^2\left(2p_2^2\alpha_1^2\alpha_2 - 2p_3^2\alpha_1^2\alpha_2 + \alpha_3\epsilon_2\right)}{\alpha_2^3}. \end{aligned} \tag{7.9}$$

With the help of eqn (7.9), along with eqn (7.8) and under consideration of Family 5, the following mathematical structure is found for $\phi(\xi)$:

$$\phi(\xi) = 2\arctan\left(\frac{\tanh\left(\frac{\kappa_1\sqrt{(p_2+p_3)(p_2-p_3)}}{2} + \frac{\xi\sqrt{(p_2+p_3)(p_2-p_3)}}{2}\right)\sqrt{(p_2+p_3)(p_2-p_3)}}{p_2-p_3}\right). \tag{7.10}$$

Thus, the analytic solution of eqn (7.1) is presented as

$$\begin{aligned} v(x,y,t) &= G_0 \\ &+ \frac{(p_2\alpha_1 - p_3\alpha_1)\tanh\left(\frac{\kappa_1\sqrt{(p_2+p_3)(p_2-p_3)}}{2} + \frac{(-t\alpha_3+x\alpha_1+y\alpha_2)\sqrt{(p_2+p_3)(p_2-p_3)}}{2}\right)\sqrt{(p_2+p_3)(p_2-p_3)}}{p_2-p_3} \\ &+ \frac{(p_2\alpha_1+p_3\alpha_1)(p_2-p_3)}{\tanh\left(\frac{\kappa_1\sqrt{(p_2+p_3)(p_2-p_3)}}{2} + \frac{(-t\alpha_3+x\alpha_1+y\alpha_2)\sqrt{(p_2+p_3)(p_2-p_3)}}{2}\right)\sqrt{(p_2+p_3)(p_2-p_3)}}, \end{aligned} \tag{7.11}$$

where $G_0,\ \alpha_1,\ p_2,\ \alpha_2,\ \alpha_3,\ p_3$, and κ_1 are arbitrary constants.

Case 2:

$$G_1 = 0,\ \epsilon_1 = \frac{\alpha_1^2\left(p_1^2\alpha_1^2\alpha_2+p_2^2\alpha_1^2\alpha_2-p_3^2\alpha_1^2\alpha_2+2\alpha_3\epsilon_2\right)}{\alpha_2^3},\ H_1 = (p_2+p_3)\alpha_1. \tag{7.12}$$

Family 1 being taken into consideration

$$\phi(\xi) = -2\arctan\left(\frac{\tan\left(\frac{\kappa_2\sqrt{-p_1^2-p_2^2+p_3^2}}{2}+\frac{\xi\sqrt{-p_1^2-p_2^2+p_3^2}}{2}\right)\sqrt{-p_1^2-p_2^2+p_3^2}-p_1}{p_2-p_3}\right), \tag{7.13}$$

where $G_0,\ p_1,\ \alpha_1,\ p_2,\ \alpha_2,\ \alpha_3,\ p_3,$ and κ_2 are free parameters.

$$v(x,y,t) = G_0 - \frac{(p_2\alpha_1+p_3\alpha_1)(p_2-p_3)}{-\tan\left(-\frac{\kappa_2\sqrt{-p_1^2-p_2^2+p_3^2}}{2}-\frac{(-t\alpha_3+x\alpha_1+y\alpha_2)\sqrt{-p_1^2-p_2^2+p_3^2}}{2}\right)\sqrt{-p_1^2-p_2^2+p_3^2}-p_1}. \tag{7.14}$$

Family 2 being taken into consideration

$$\phi(\xi) = 2\arctan\left(\frac{p_1}{p_2-p_3}+\frac{\sqrt{p_1^2+p_2^2-p_3^2}\tanh\left(\frac{\sqrt{p_1^2+p_2^2-p_3^2}\,(\xi+\kappa_3)}{2}\right)}{p_2-p_3}\right) \tag{7.15}$$

$$v(x,y,t) = G_0 + \frac{p_2\alpha_1+p_3\alpha_1}{\frac{p_1}{p_2-p_3}+\frac{\sqrt{p_1^2+p_2^2-p_3^2}\tanh\left(\frac{\sqrt{p_1^2+p_2^2-p_3^2}\,(-t\alpha_3+x\alpha_1+y\alpha_2+\kappa_3)}{2}\right)}{p_2-p_3}}, \tag{7.16}$$

where $G_0,\ p_1,\ \alpha_1,\ p_2,\ \alpha_2,\ \alpha_3,\ p_3,$ and κ_3 are arbitrary parameters.

Case 3:

$$H_1 = 0,\ G_1 = (p_2-p_3)\alpha_1,\ \epsilon_1 = \frac{\alpha_1^2\left(p_1^2\alpha_1^2\alpha_2+p_2^2\alpha_1^2\alpha_2-p_3^2\alpha_1^2\alpha_2+2\alpha_3\epsilon_2\right)}{\alpha_2^3}. \tag{7.17}$$

Family 1 being taken into consideration

$$\phi(\xi) = -2\arctan\left(\frac{\tan\left(\frac{\kappa_4\sqrt{-p_1^2-p_2^2+p_3^2}}{2}+\frac{\xi\sqrt{-p_1^2-p_2^2+p_3^2}}{2}\right)\sqrt{-p_1^2-p_2^2+p_3^2}-p_1}{p_2-p_3}\right), \tag{7.18}$$

where $G_0,\ p_1,\ \alpha_1,\ p_2,\ \alpha_2,\ \alpha_3,\ p_3$, and κ_4 are free constants.

$$v(x,y,t) = G_0 - \frac{(p_2\alpha_1 - p_3\alpha_1)\left(-\tan\left(-\frac{\kappa_4\sqrt{-p_1^2-p_2^2+p_3^2}}{2} - \frac{(-t\alpha_3+x\alpha_1+y\alpha_2)\sqrt{-p_1^2-p_2^2+p_3^2}}{2}\right)\sqrt{-p_1^2-p_2^2+p_3^2} - p_1\right)}{p_2 - p_3}. \tag{7.19}$$

Family 2 being taken into consideration

$$\phi(\xi) = 2\arctan\left(\frac{p_1}{p_2 - p_3} + \frac{\sqrt{p_1^2+p_2^2-p_3^2}\tanh\left(\frac{\sqrt{p_1^2+p_2^2-p_3^2}\,(\xi+\kappa_5)}{2}\right)}{p_2 - p_3}\right) \tag{7.20}$$

$$v(x,y,t) = G_0 + (p_2\alpha_1 - p_3\alpha_1)\left(\frac{p_1}{p_2 - p_3} + \frac{\sqrt{p_1^2+p_2^2-p_3^2}\tanh\left(\frac{\sqrt{p_1^2+p_2^2-p_3^2}\,(-t\alpha_3+x\alpha_1+y\alpha_2+\kappa_5)}{2}\right)}{p_2 - p_3}\right), \tag{7.21}$$

where $G_0,\ p_1,\ \alpha_1,\ p_2,\ \alpha_2,\ \alpha_3,\ p_3$, and κ_5 are arbitrary constants.

Case 4:

$$p_2 = p_3,\ G_1 = 0,\ \alpha_3 = -\frac{\alpha_2\left(p_1^2\alpha_1^4 - \alpha_2^2\epsilon_1\right)}{2\alpha_1^2\epsilon_2},\ H_1 = 2\alpha_1 p_3. \tag{7.22}$$

Family 16 being taken into consideration

$$\phi(\xi) = 2\arctan\left(\frac{e^{p_1(\kappa_6+\xi)} - p_3}{p_1}\right) \tag{7.23}$$

$$v(x,y,t) = G_0 + \frac{2\alpha_1 p_3 p_1}{e^{\kappa_6 p_1 + \left(\frac{t\alpha_2\left(p_1^2\alpha_1^4 - \alpha_2^2\epsilon_1\right)}{2\alpha_1^2\epsilon_2} + x\alpha_1 + y\alpha_2\right)p_1} - p_3}, \tag{7.24}$$

where $G_0,\ p_1,\ \alpha_1,\ \alpha_2,\ p_3, \kappa_6,\ \epsilon_2$, and ϵ_1 are arbitrary constants.

Case 5:

$$H_1 = 0,\ p_2 = -p_3,\ \alpha_1 = -\frac{G_1}{2p_3}. \tag{7.25}$$

Family 12 being taken into consideration

$$\phi(\xi) = -2\arctan\left(\frac{e^{p_1(\kappa_7+\xi)}p_1}{-1+e^{p_1(\kappa_7+x)}p_3}\right) \tag{7.26}$$

$$v(x,y,t) = G_0 - \frac{G_1 e^{\frac{\kappa_7 p_1 + \left(-t\alpha_3 - \frac{xG_1}{2p_3} + y\alpha_2\right) p_1}{p_1}}}{-1 + e^{\frac{\kappa_7 p_1 + \left(-t\alpha_3 - \frac{xG_1}{2p_3} + y\alpha_2\right) p_1}{p_3}}}, \tag{7.27}$$

where $G_0,\ G_1,\ p_1,\ \alpha_2,\ \alpha_3,\ p_3$, and κ_7 are arbitrary parameters.

Case 6:

$$p_3 = -p_1,\ G_1 = (p_1 + p_2)\alpha_1,\ \epsilon_1 = \frac{\alpha_1^2\left(p_2^2\alpha_1^2\alpha_2 + 2\epsilon_2\alpha_3\right)}{\alpha_2^3},\ H_1 = 0, \tag{7.28}$$

Family 11 being taken into consideration

$$\phi(\xi) = 2\arctan\left(\frac{p_2 + e^{p_2\xi + \kappa_8 p_2} - p_1}{e^{p_2\xi + \kappa_8 p_2} - p_1 - p_2}\right) \tag{7.29}$$

$$v(x,y,t) = G_0 + \frac{(p_1\alpha_1 + p_2\alpha_1)\left(p_2 + e^{\kappa_8 p_2 + p_2(-t\alpha_3 + x\alpha_1 + y\alpha_2)} - p_1\right)}{e^{\kappa_8 p_2 + p_2(-t\alpha_3 + x\alpha_1 + y\alpha_2)} - p_1 - p_2}, \tag{7.30}$$

where $G_0,\ p_1,\ \alpha_1,\ \alpha_2,\ \alpha_3, \kappa_8$, and p_2 are free constants.

Case 7:

$$p_2 = p_1,\ G_1 = 0,\ \epsilon_2 = -\frac{\alpha_2\left(p_1^2\alpha_1^4 - \alpha_2^2\epsilon_1\right)}{2\alpha_1^2\alpha_3},\ p_3 = p_1,\ H_1 = 2p_1\alpha_1. \tag{7.31}$$

Family 7 being taken into consideration

$$\phi(\xi) = 2\arctan\left(e^{(\kappa_9 + \xi)p_1} - 1\right) \tag{7.32}$$

$$v(x,y,t) = G_0 + \frac{2p_1\alpha_1}{e^{\kappa_9 p_1 + p_1(-t\alpha_3 + x\alpha_1 + y\alpha_2)} - 1}, \tag{7.33}$$

where $G_0,\ p_1,\ \alpha_1,\ \alpha_2,\ \alpha_3$, and κ_9 are arbitrary parameters.

Case 8:

$$\begin{aligned} &p_1 = p_3,\ G_1 = 0,\\ &\epsilon_2 = \frac{\left(p_2^4\alpha_2^2\epsilon_1 + 4p_2^3p_3\alpha_2^2\epsilon_1 + 6p_2^2p_3^2\alpha_2^2\epsilon_1 + 4p_2\,p_3^3\alpha_2^2\epsilon_1 + p_3^4\alpha_2^2\epsilon_1 - p_2^2H_1^4\right)\alpha_2}{2\alpha_3 H_1^2(p_2+p_3)^2},\ \alpha_1 = \frac{H_1}{p_2+p_3}. \end{aligned} \tag{7.34}$$

Family 10 being taken into consideration

$$\phi(\xi) = 2\arctan\left(\frac{(p_2+p_3)e^{(\kappa_{10}+\xi)p_2} + 1}{-1 + (p_2 - p_3)e^{(\kappa_{10}+\xi)p_2}}\right) \tag{7.35}$$

$$\begin{aligned} v(x,y,t) &= G_0\\ &+ \frac{H_1\left(-1 + p_2\, e^{\kappa_{10}p_2 + \left(-t\alpha_3 + \frac{xH_1}{p_2+p_3} + y\alpha_2\right)p_2} - p_3\, e^{\kappa_{10}p_2 + \left(-t\alpha_3 + \frac{xH_1}{p_2+p_3} + y\alpha_2\right)p_2}\right)}{p_2\, e^{\kappa_{10}p_2 + \left(-t\alpha_3 + \frac{xH_1}{p_2+p_3} + y\alpha_2\right)p_2} + p_3\, e^{\kappa_{10}p_2 + \left(-t\alpha_3 + \frac{xH_1}{p_2+p_3} + y\alpha_2\right)p_2} + 1}, \end{aligned} \tag{7.36}$$

where $G_0,\ H_1\ p_2,\ \alpha_3,\ p_3,\ \alpha_2$, and κ_{10} are free parameters.

Case 9:

$$p_2 = -\beta p_1,\ p_3 = \beta p_1,\ G_1 = -2p_1\beta\alpha_1,\ H_1 = 0,\ \epsilon_1 = \frac{\alpha_1^2\left(p_1^2\alpha_1^2\alpha_2+2\alpha_3\epsilon_2\right)}{\alpha_2^3}. \tag{7.37}$$

Family 8 being taken into consideration

$$\phi(\xi) = -2\arctan\left(\frac{e^{(\kappa_{11}+\xi)p_1}}{-1+e^{(\kappa_{11}+\xi)p_1}\beta}\right) \tag{7.38}$$

$$v(x,y,t) = G_0 + \frac{2p_1\beta\alpha_1 e^{\kappa_{11}p_1+(-t\alpha_3+x\alpha_1+y\alpha_2)p_1}}{-1+e^{\kappa_{11}p_1+(-t\alpha_3+x\alpha_1+y\alpha_2)p_1}\beta}, \tag{7.39}$$

where $G_0,\ \beta,\ p_1,\ \alpha_1,\ \kappa_{11}\ \alpha_2,\ \alpha_3$, and κ_{11} are free constants.

Case 10:

$$p_1 = 0,\ p_3 = 0,\ G_1 = p_2\alpha_1,\ \alpha_3 = -\frac{\alpha_2\left(4p_2^2\alpha_1^4-\alpha_2^2\epsilon_1\right)}{2\alpha_1^2\epsilon_2},\ H_1 = p_2\alpha_1. \tag{7.40}$$

Family 17 being taken into consideration

$$\phi(\xi) = \arctan\left(\frac{e^{2p_2\xi+2\kappa_{12}p_2}-1}{e^{2p_2\xi+2\kappa_{12}p_2}+1}, \frac{2e^{p_2\xi+\kappa_{12}p_2}}{e^{2p_2\xi+2\kappa_{12}p_2}+1}\right) \tag{7.41}$$

$$\begin{aligned} v(x,y,t) = G_0 &+ p_2\alpha_1\tan\left(\frac{\arctan\left(\frac{e^{\frac{p_2\left(4p_2^2t\alpha_1^4\alpha_2-t\alpha_2^3\epsilon_1+2x\alpha_1^3\epsilon_2+2y\alpha_2\alpha_1^2\epsilon_2+2\kappa_{11}\alpha_1^2\epsilon_2\right)}{\alpha_1^2\epsilon_2}}-1}{2e^{\frac{p_2\left(4p_2^2t\alpha_1^4\alpha_2-t\alpha_2^3\epsilon_1+2x\alpha_1^3\epsilon_2+2y\alpha_2\alpha_1^2\epsilon_2+2\kappa_{11}\alpha_1^2\epsilon_2\right)}{2\alpha_1^2\epsilon_2}}}\right)}{2}\right) \\ &+ p_2\alpha_1\cot\left(\frac{\arctan\left(\frac{e^{\frac{p_2\left(4p_2^2t\alpha_1^4\alpha_2-t\alpha_2^3\epsilon_1+2x\alpha_1^3\epsilon_2+2y\alpha_2\alpha_1^2\epsilon_2+2\kappa_{11}\alpha_1^2\epsilon_2\right)}{\alpha_1^2\epsilon_2}}-1}{2e^{\frac{p_2\left(4p_2^2t\alpha_1^4\alpha_2-t\alpha_2^3\epsilon_1+2x\alpha_1^3\epsilon_2+2y\alpha_2\alpha_1^2\epsilon_2+2\kappa_{11}\alpha_1^2\epsilon_2\right)}{2\alpha_1^2\epsilon_2}}}\right)}{2}\right), \end{aligned} \tag{7.42}$$

where $G_0,\ p_2,\ \alpha_1,\ \alpha_2,\ \epsilon_1, \epsilon_2$, and κ_{12} are free parameters.

7.4 Conclusion

In summary, by means of $\tan\left(\frac{\phi(\xi)}{2}\right)$-expansion method, $(2+1)$-dimensional Date–Jimbo–Kashiwara–Miwa equation has been successfully solved. In high dimensions, numerous universal phenomena might be judged on the basis of richness of the solutions, which comes from the presence of some arbitrary constants. On the other hand, rich variety of the obtained solutions and their diverse behavior depicts about the relevance of the current method on the system of such nonlinear evolutionary partial differential equations. A confrontation of our results with the already existing solutions has been done and come to a realization that the results explored in this research work are completely new. Because of the widespread adherence of the nonlinear theory to the sense of reality, more about the properties related to such solutions is worth of further study.

Conflict of Interest Statement

On behalf of all authors, the corresponding author states that there is no conflict of interest.

References

[1] L. Kaur and A. M. Wazwaz, Complex simplified Hirota's forms and Lie symmetry analysis for multiple real and complex soliton solutions of the modified KdVŰSine-Gordon equation, *Nonlinear Dynamics* 95 (2019): 2209-2215.

[2] X. Zheng and H. Wu, Orbital stability of periodic traveling wave solutions to the coupled compound KdV and MKdV equations with two components, *Mathematical Foundations of Computing* 3 (2020): 11-24.

[3] Y. Xu, X. Zheng and J. Xin, New explicit and exact traveling wave solutions of (3+1)-dimensional KP equation, *Mathematical Foundations of Computing* 4 (2021): 105-115.

[4] S. Malik, S. Kumar and A. Das, A (2+1)-dimensional combined KdV-mKdV equation: integrability, stability analysis and soliton solutions, *Nonlinear Dynamics* 107 (2022): 2689-2701.

[5] S. Shagolshem, B. Bira and S. Sil, Application of symmetry analysis to viscoelastic fluid model, *Communications in Nonlinear Science and Numerical Simulation* 125 (2023) 107417,

[6] G. Akram, N. Sajid, M. Abbas , Y. S. Hamed and K. M. Abualnaja, Optical solutions of the Date–Jimbo–Kashiwara–Miwa equation via

the extended direct algebraic method, *Journal of Mathematics* 2021 (2021): 5591016.

[7] A. Chauhan, K. Sharma and R. Arora, Lie symmetry analysis, optimal system, and generalized group invariant solutions of the $(2+1)$-dimensional Date–Jimbo–Kashiwara–Miwa equation, *Mathematical Methods in the Applied Sciences* 43 (2020): 8823-8840.

[8] A. R. Adem, Y. Yildirim and E. Yaşar, Complexiton solutions and soliton solutions: $(2+1)$-dimensional Date–Jimbo–Kashiwara–Miwa equation, *Pramana* 92 (2019): 36.

[9] K. K. Ali, M. S. Mehanna and A. M. Wazwaz, Analytical and numerical treatment to the $(2+1)$-dimensional Date–Jimbo–Kashiwara–Miwa equation, *Nonlinear Engineering* 10 (2021): 187-200.

[10] F. Guo and J. Lin, Interaction solutions between lump and stripe soliton to the $(2+1)$-dimensional Date–Jimbo–Kashiwara–Miwa equation, *Nonlinear Dynamics* 96 (2019): 1233-1241.

[11] Y.-H. Wang, H. Wang, C. Temuer, Lax pair, conservation laws, and multi-shock wave solutions of the DJKM equation with Bell polynomials and symbolic computation, *Nonlinear Dynamics* 78 (2014): 1101-1107.

[12] L. Cheng, Y. Zhang, M. J. Lin, Lax pair and lump solutions for the $(2+1)$-dimensional DJKM equation associated with bilinear Bäcklund transformations, *Analysis and Mathematical Physics* 9 (2019): 1741-1752.

[13] H. F. Ismael, H. Bulut, C. Park, M. S. Osman, M-lump, N-soliton solutions, and the collision phenomena for the $(2+1)$-dimensional Date–Jimbo–Kashiwara–Miwa equation, *Results in Physics* 19 (2020): 103329.

[14] S. Kumar and A. Kumar, Dynamical structures of solitons and some new types of exact solutions for the (2+1)-dimensional DJKM equation using Lie symmetry analysis, *Modern Physics Letters B* 34 (2020): 2150015.

[15] A. M. Wazwaz, A $(2+1)$-dimensional time-dependent Date–Jimbo–Kashiwara–Miwa equation: Painlevé integrability and multiple soliton solutions, *Computers and Mathematics with Applications* 79 (2020): 1145-1149

[16] M. F. Aghdaei, J. Manafian and M. Zadahmad, Analytic study of sixth-order thin film equation by $\tan\left(\frac{\phi(\xi)}{2}\right)$-expansion method, *Optical and Quantum Electronics* 48 (2016): 410.

[17] J. Manafian and M. Lakestani, A. Bekir, Study of the analytical treatment of the $(2 + 1)$-dimensional Zoomeron, the Duffing and the SRLW equations via a new analytical approach, *International Journal of Applied and Computational Mathematics* 2 (2016); 243-268.

[18] J. Manafian and M. Lakestani, Abundant soliton solutions for the Kundu-Eckhaus equation via $\tan\left(\frac{\phi(\xi)}{2}\right)$-expansion method, *Optik* 127 (2016): 5543-5551.

[19] N. Ahmed, A. Irshad, S. T. M. Din and U. Khan, Exact solutions of perturbed nonlinear Schrödinger's equation with Kerr law nonlinearity by improved $\tan\left(\frac{\phi(\xi)}{2}\right)$-expansion method, *Optical and Quantum Electronics* 50 (2018): 45.

[20] P. Verma and L. Kaur, Solitary wave solutions for $(1 + 2)$-dimensional nonlinear Schrödinger equation with dual power law nonlinearity, *International Journal of Applied and Computational Mathematics* 5 (2019): 128.

8

Unveiling Dynamics of Solutions for Variable Killing Rate Brain Tumor Model With Lie Symmetry Analysis

Lakhveer Kaur[1], K. Murugesan[2], and Aman Fatima[1]

[1]Department of Mathematics,
Jaypee Institute of Information Technology, India
[2]Department of Mathematics, National Institute of Technology, India
E-mail: lakhveer712@gmail.com; murugu@nitt.edu;
amanfatima204@gmail.com

Abstract

The complexities of brain tumor dynamics demand advanced mathematical techniques to unfold underlying patterns and optimize treatment strategies, and thus mathematical models along with their extensive analysis have recently been employed in the study of both untreated and cured brain tumors. This course of study involves the examination of the model representing a brain tumor with variable killing rates, being expressed in the form of a non-linear PDE using Lie symmetry analysis, and, consecutively, the exploration of infinitesimals is utilized to reveal the corresponding reduced ODEs in an ordered way, leading to the formulation of solutions to the considered model from the perspective of similarity variables. Additionally, some solutions are found with the help of the power series method and then after discussing the convergence of these power series solutions to establish the efficiency and efficacy of extracted solutions, which might have some practical implications in real-life situations.

Keywords: Brain tumor model, Lie symmetry method, exact solutions, power series

8.1 Introduction

Over the past few decades, monitoring and growth of brain tumors has captured a great deal of interest. The interdisciplinary fields of research become considerably more obsessed with these kinds of real-world problems. A subset of bio-mathematical modeling is the modeling of brain tumors, in which mathematical models are used to investigate the tumor growth phenomena, which is exclusively dependent on response diffusion. Cancer cells multiply and spread across a brain tumor's dynamic structure until they kill healthy brain cells by depriving them of nutrients and space. For brain tumor diagnosis, measurements of apparent diffusion coefficient (ADC) in brain tumors allow doctors to categorize tumor type.

Mathematical modeling of brain tumors makes use of computational techniques and mathematical equations to mimic and understand the development and activity of brain tumors. These models can provide insight into the development, progression, and response to treatment of cancers. Here are some key models used in mathematical modeling of brain tumors: tumor growth models, reaction–diffusion models, agent-based models, biomechanical models, treatment response models, imaging and data integration, etc.

Overall, mathematical modeling of brain tumors provides a quantitative framework to understand tumor growth dynamics, optimize treatment strategies, and explore potential therapeutic interventions. Here, the main concern of the type of model is reaction–diffusion model.

Many of the nonlinear problems can be modeled into a differential equation, likewise here we have a nonlinear differential equation, explicitly a PDE, which will be converted into an ODE by using Lie symmetry analysis, and further we will find solutions in different cases. The challenge of solving nonlinear differential equations has drawn a lot of interest because nonlinear problems frequently arise in real-life scenarios. Additionally, it may be possible to attain physical relevance of the extracted solutions, followed by case studies of applications in the physical and biological sciences. Due to the richness of the data produced and potential applications, the absolute examination of solutions to nonlinear partial differential equations (NLPDEs) occurs to be the main tool for studying numerous physical conditions. Many well-known and reliable strategies have recently been successfully applied to parse NLPDEs with constant and variable coefficients.

Further effective approaches for solving NLPDEs include the generalized $\left(\frac{G'}{G}\right)$ expansion method [1], sine-cosine analysis [2], F-expansion analysis [3], extended tanh analysis [4, 5], ansatz methods [6], exp $(\varphi\,(\xi))$ expansion function [7], and few more [8, 9]. These methods lead to a unifying concept in the systematic creation of accurate solutions to various NLPDEs. Because these methods can involve lengthy and expensive calculations, software such as Maple and Mathematica can be used to simplify calculations. Furthermore, physical relevance of these extracted solutions may be realized, which will be followed by case studies of applications in the physical and biological sciences.

8.2 Lie Symmetry Analysis of Brain Tumor Model

Lie symmetry analysis [10, 11] can be applied to reaction–diffusion brain models to classify the symmetries present in the system. By identifying the symmetries, we can gain insights into the invariance properties of the model, revealing which transformations leave the model's equations unchanged. This classification can help identify fundamental features of the model and provide a better understanding of its underlying structure. The reaction–diffusion equation is a mathematical model that describes how substances diffuse and interact with each other in a system.

In certain mathematical models, we study reaction–diffusion equation to simulate and study the growth and spread of brain tumors. Here, the reaction term represents the cell proliferation and growth of tumor cells, while the diffusion term describes the movement of cells within the brain tissue. By incorporating parameters related to tumor cell proliferation rates, migration rates, and the diffusivity of cells, the reaction–diffusion equation can provide insights into the spatial and temporal dynamics of brain tumor growth.

The tumor cells are capable to spread fast by the killing of healthy cells under the absence of oxygen and nutrients. In this case, we have been using the very popular model [12] with the help of the Burgers equation, which is based on the process of reaction–diffusion coupled with the killing rate, which depends upon therapy. Consider the mathematical model [12], which is known as Burgers' equation, given by

$$\frac{\partial n\left(r,t\right)}{\partial t}=D\frac{1}{r^{2}}\frac{\partial}{\partial r}\left(\frac{\partial n\left(r,t\right)}{\partial r}\right)+pn\left(r,t\right)-\ k_{t}n\left(r,t\right), \tag{8.1}$$

where n is the concentration of tumor cells, which depends on radial distance r and time t, D is the diffusion coefficient, which measures the invasiveness of the cells of glioblastoma multiforme, p is the proliferation rate, which tells the growth and division of tumor cells, and k_t is the killing rate, which is therapy dependent.

By performing rescaling of variables r and t as $t = (p - k_t)T$, $R = (|p - k_t| D)^2 r$ and reverting the upper case variables to lower case variables, we can rewrite eqn (8.1) as

$$\frac{\partial n}{\partial t} = \frac{\partial^2 n}{\partial r^2} + \frac{2}{r}\frac{\partial n}{\partial r} + n. \tag{8.2}$$

By substitution $u(x,t) = r\,n(x,t)$, $x = r$, in eqn (8.2) we get

$$\frac{\partial u}{\partial t} = \frac{\partial^2 u}{\partial x^2} + u. \tag{8.3}$$

Consider the more general and interesting case of eqn (8.1) for single cell type [13], which is given by

$$\frac{\partial^2 u}{\partial x^2} - K(u)\;u - \frac{\partial u}{\partial t} = 0, \tag{8.4}$$

where $K(u)$ is the removal rate of the tumor cells. The model considered here is the function of the time and position of the cancerous cells, which is an even more genuine assumption. Now we will discuss Lie symmetry analysis for the solutions of the tumor eqn (8.4), to study the exact solutions of the model with different cases of $K(u)$.

We will consider the two general cases for the $K(u)$:

Case 1: $K(u) = u^m$.

Then eqn (8.4) is rewritten as

$$\frac{\partial^2 u}{\partial x^2} - u^{m+1} - \frac{\partial u}{\partial t} = 0. \tag{8.5}$$

The corresponding symmetries for $K(u) = u^m$ are successfully furnished as:

$$\xi = \frac{1}{2}xc_1, \tau = c_2 t + c_2,\ \eta = \frac{-c_1 u}{m},$$

and then vector fields w.r.t. coefficients c_1 and c_2 respectively are discovered as follows:

$$V_1 = \frac{x}{2}\frac{\partial}{\partial x} + t\frac{\partial}{\partial t} - \frac{u}{n}\frac{\partial}{\partial u},\quad V_2 = \frac{\partial}{\partial t}.$$

Now, reduced ODE for eqn (8.5) w.r.t. the vector field V_1 is

$$4\xi\left(F''(\xi)\right) + (2+\xi)\left(F'(\xi)\right) - F(\xi)^{m+1} + \frac{F(\xi)}{m} = 0. \tag{8.6}$$

Using another vector field V_2, the corresponding ODE for eqn (8.5) is

$$F''(\xi) - F(\xi)^{m+1} = 0. \tag{8.7}$$

Thereafter, we will discuss the solutions of ODE eqn (8.6) and (8.7) in Section 3.

Case 2: $K(u) = \log u$.

Then eqn (8.4) is rewritten as

$$\frac{\partial^2 u}{\partial x^2} - u \log u - \frac{\partial u}{\partial t} = 0. \tag{8.8}$$

The corresponding symmetries for $K(u) = \log u$ are successfully furnished as

$$\xi = 0, \tau = c_1, \eta = c_2 u e^{-t},$$

and the following are the Lie operators V_1 and V_2 w.r.t. coefficients c_1 and c_2, respectively:

$V_1 = \frac{\partial}{\partial t}$, $V_2 = ue^{-t}\frac{\partial}{\partial u}$. Next, we consider $\langle V_1 + \mu V_2 \rangle = \frac{\partial}{\partial t} + \mu u e^{-t}\frac{\partial}{\partial u}$, for reducing eqn (8.8) into corresponding ODE, which is written as follows:

$$F''(\xi) - F(\xi)\log(F(\xi)) = 0. \tag{8.9}$$

8.3 Establishing Exact Solutions via Power Series Method

In this section, we will discuss the solutions of the different ODEs obtained in Section 2 with the help of the powerful tool of power series [14]. As we know, the solution of nonlinear systems is not an easy task, and even sometimes they are unable to solve. Lie symmetry transformations were used to produce reduced ordinary differential equations. We will look at the obtained nonlinear ODEs in this section.

In general context, applying the basic functions and integrals does not allow us to derive precise explicit solutions for these nonlinear ODEs (obtained in Section 2). However, ODEs, including numerous challenging differential equations with nonconstant coefficients, can be resolved using the power series. Therefore, we use the power series solution method to obtain the solutions of the reduced differential equations.

8.3.1 Exact solution of eqn (2.6)

We have obtained the ODE in eqn (8.6) in Section 2. Now we are finding the solution using the power series solution method. For $m = 1$, we get

$$4\xi \left(F''(\xi)\right) + (2+\xi)\left(F'(\xi)\right) - F(\xi)^2 + F(\xi) = 0. \tag{8.10}$$

Assuming that the power series is structured as follows:

$$F(\xi) = \sum_{n=0}^{\infty} c_n \xi^n, \tag{8.11}$$

we have

$$F'(\xi) = \sum_{n=0}^{\infty} (n+1) c_{n+1} \xi^n \tag{8.12}$$

$$F''(\xi) = \sum_{n=0}^{\infty} (n+1)(n+2) c_{n+2} \xi^n \tag{8.13}$$

Now substituting the values of $F(\xi)$, $F'(\xi)$, and $F''(\xi)$ in eqn (8.10) we get

$$4\left(\sum_{n=0}^{\infty} (n+1)(n+2) c_{n+2} \xi^{n+1}\right) + (2+\xi)\left(\sum_{n=0}^{\infty} (n+1) c_{n+1} \xi^n\right)$$
$$\left(\sum_{n=0}^{\infty} n c_n \xi^n\right)^2 + \left(\sum_{n=0}^{\infty} (n+1) c_{n+1} \xi^n\right) = 0. \tag{8.14}$$

On rearranging eqn (8.14) we get

$$4\left(2c_1 + c_0 - c_0^2\right) + \sum_{n=1}^{\infty} \left(4n(n+1)c_{n+1} + nc_n - \sum_{k=0}^{n} c_k c_{n-k} + c_n\right) \xi^n = 0 \tag{8.15}$$

From eqn (8.15) comparing the coefficients for different powers of ξ, we get

$$c_1 = \frac{1}{2}\left(c_0^2 - c_0\right) \tag{8.16}$$

$$c_{n+1} = \frac{1}{4n(n+1)} \left(\sum_{k=0}^{n} c_k c_{n-k} - nc_n - c_n\right), \quad n \geq 1$$

where ($c_i,\ i = 1, 2\ldots$) are arbitrary constants.

For $n = 1,\ 2$ we obtain

$$c_2 = \frac{1}{4}(\,c_0c_1 - c_1),\ c_3 = \frac{1}{24}\left(\,2c_0c_2 + {c_1}^2 - 3c_2\right).$$

Hence, the power series solution of eqn (8.6) can be expressed as

$$\begin{aligned} F\left(\xi\right) &= c_0 + c_1\xi + \sum_{n=1}^{\infty} c_{n+1}\xi^{n+1} \\ &= c_0 + \frac{1}{2}\left({c_0}^2 - c_0\right)\xi \\ &\quad + \sum_{n=1}^{\infty}\frac{1}{4n\left(n+1\right)}\left(\sum_{k=0}^{n} c_kc_{n-k} - nc_n - c_n\right)\xi^{n+1}. \end{aligned}$$

Thus, the exact solution of eqn (8.5) for $m = 1$ is

$$\begin{aligned} u(x,t) &= c_0t^{-1} + \frac{1}{2}\left({c_0}^2 - c_0\right)\left(\frac{x}{t}\right)^2 \\ &\quad + \sum_{n=1}^{\infty}\frac{1}{4n\left(n+1\right)}\left(\sum_{k=0}^{n} c_kc_{n-k} - nc_n - c_n\right)\left(\frac{x}{t}\right)^{2(n+1)}. \end{aligned}$$

8.3.2 Exact solution of eqn (2.7)

We have considered ODE in eqn (8.7) for $m = 2$. Then accordingly we have

$$F''\left(\xi\right) - F\left(\xi\right)^2 = 0. \tag{8.17}$$

Let $F\left(\xi\right) = \sum_{n=0}^{\infty} c_n\xi^n$ be the power series solution for eqn (8.17). Then we have

$$F'\left(\xi\right) = \sum_{n=0}^{\infty}\left(n+1\right)c_{n+1}\xi^n,\ F''\left(\xi\right) = \sum_{n=0}^{\infty}\left(n+1\right)\left(n+2\right)c_{n+2}\xi^n.$$

Now substituting the values of $F\left(\xi\right)$ and $F''\left(\xi\right)$ in eqn (8.17) we obtain

$$\left(\sum_{n=0}^{\infty}\left(n+1\right)\left(n+2\right)c_{n+2}\xi^n\right) - \left(\sum_{n=0}^{\infty} c_n\xi^n\right)^2 = 0. \tag{8.18}$$

After rearranging eqn (8.18), we get

$$\left(2c_2 - {c_0}^3\right) + \sum_{n=1}^{\infty}\left(\left(n+1\right)\left(n+2\right)c_{n+2} - \sum_{k=0}^{n}\sum_{i=0}^{k} c_i c_{k-i} c_{n-k}\right)\xi^n = 0. \tag{8.19}$$

On comparing the coefficients for different powers of ξ, we get

$$c_2 = \frac{{c_0}^3}{2}, \quad c_{n+2} = \frac{1}{\left(n+1\right)\left(n+2\right)}\left(\sum_{k=0}^{n}\sum_{i=0}^{k} c_i c_{k-i} c_{n-k}\right), \quad n \geq 1 \tag{8.20}$$

where (c_i, $i = 1, 2\ldots$) are arbitrary constants.

Putting $n = 1, 2$ in eqn (8.20) we obtain

$$c_3 = \frac{c_1 {c_0}^2}{2}, c_4 = \frac{1}{4}\left({c_0}^2 c_2 + 2c_0 {c_1}^2\right).$$

Hence, the power series solution of eqn (8.7) can be expressed as

$$\begin{aligned} F\left(\xi\right) &= c_0 + c_1\xi + c_2\xi^2 + \sum_{n=1}^{\infty} c_{n+2}\xi^{n+2}, \\ &= c_0 + c_1\xi + \frac{{c_0}^3}{2}\xi^2 \\ &\quad + \sum_{n=1}^{\infty}\frac{1}{\left(n+1\right)\left(n+2\right)}\left(\sum_{k=0}^{n}\sum_{i=0}^{k} c_i c_{k-i} c_{n-k}\right)\xi^{n+2}. \end{aligned}$$

Thus, the exact solution of eqn (8.5) for $m = 2$ is

$$\begin{aligned} u\left(x,t\right) &= c_0 + c_1 x + \frac{{c_0}^3}{2}x^2 \\ &\quad + \sum_{n=1}^{\infty}\frac{1}{\left(n+1\right)\left(n+2\right)}\left(\sum_{k=0}^{n}\sum_{i=0}^{k} c_i c_{k-i} c_{n-k}\right)x^{n+2}. \end{aligned}$$

But because it lacks its temporal component, this kind of solution does not hold much of our interest.

8.3.3 Exact solution via traveling wave transformation

In this subsection, we are going to explore the wave transformation $\xi = \kappa x - \omega t$; $u\left(x,t\right) = F\left(\xi\right)$ to establish some new solutions for the considered brain tumor model as eqn (8.5) with $m = 2$; as a result, the subsequent ODE is furnished to carry out further analysis:

$$\kappa^2 F''(\xi) + \omega F'(\xi) - F(\xi)^3 = 0, \tag{8.21}$$

where $\kappa \neq 0,\ \omega \neq 0,\ \kappa$ and ω are arbitrary constants.

Now apply the power series method in eqn (8.21) along with the following power series

$$F(\xi) = \sum_{n=0}^{\infty} c_n \xi^n. \tag{8.22}$$

Differentiating eqn (8.22) up to two times, we obtain

$$F'(\xi) = \sum_{n=0}^{\infty} (n+1)\, c_{n+1}\xi^n, \quad F''(\xi) = \sum_{n=0}^{\infty} (n+1)(n+2)\, c_{n+2}\xi^n.$$

Substituting the values of $F(\xi)$, $F'(\xi)$, and $F''(\xi)$ in eqn (8.21) we get

$$\kappa^2 \left(\sum_{n=0}^{\infty} (n+1)\ (n+2) c_{n+2}\xi^n \right) + \omega \left(\sum_{n=0}^{\infty} (n+1) c_{n+1}\xi^n \right) \tag{8.23}$$
$$- \left(\sum_{n=0}^{\infty} c_n \xi^n \right)^3 = 0.$$

On comparing the coefficients for different powers of ξ in eqn (8.23), we get

$$c_2 = \frac{c_0^3 - \omega c_1}{2\,\kappa^2},\quad c_{n+2} = \frac{1}{\kappa^2 (n+1)(n+2)} \tag{8.24}$$
$$\left(\sum_{k=0}^{n} \sum_{i=0}^{k} c_i c_{k-i} c_{n-k} - \omega (n+1)\, c_{n+1} \right),\quad n \geq 1$$

where (c_i, $i = 1, 2 \ldots$) are arbitrary constants.

Putting $n = 1,\ 2$ in eqn (8.24), we obtain

$$c_3 = \frac{1}{6\,\kappa^2}\left(3c_1 c_0^2 - 2\omega c_2\right),$$
$$c_4 = \frac{1}{12\,\kappa^2}\left(c_1 c_0^2 + 3c_0 c_1^2 + 2c_2 c_0^2 - 3\omega c_3\right).$$

Hence, the power series solution of eqn (8.21) can be expressed as

$$\begin{aligned} F(\xi) &= c_0 + c_1\xi + c_2\xi^2 + \sum_{n=1}^{\infty} c_{n+2}\xi^{n+2}, \\ &= c_0 + c_1\xi + \frac{{c_0}^3 - \omega c_1}{2\,\kappa^2}\,\xi^2 \\ &\quad + \sum_{n=1}^{\infty} \frac{1}{(n+1)(n+2)} \left(\sum_{k=0}^{n}\sum_{i=0}^{k} c_i c_{k-i} c_{n-k} \right) \xi^{n+2}. \end{aligned}$$

Therefore, the exact solution of eqn (8.4) for $m = 2$ is

$$\begin{aligned} u(x,t) &= c_0 + c_1(\kappa x - \omega t) + \frac{{c_0}^3 - \omega c_1}{2\kappa^2}(\kappa x - \omega t)^2 \\ &\quad + \sum_{n=1}^{\infty} \frac{1}{(n+1)(n+2)} \left(\sum_{k=0}^{n}\sum_{i=0}^{k} c_i c_{k-i} c_{n-k} \right) (\kappa x - \omega t)^{n+2}. \end{aligned} \tag{8.25}$$

8.4 Convergence of Power Series Solution

In this section, we will discuss the convergence of the power series solution obtained in the previous section. The convergence of the power series method tells us about the accuracy of the solution obtained; if the solution obtained is convergent, then our solution is convergent, otherwise it is divergent, i.e., it is unbounded. In particular, we have taken up eqn (8.25) for elaborating the process of discovering its convergence and then after in the similar fashion, the convergence of other found power series solutions can be analyzed.

Now let us show the convergence [14] of the power series solution eqn (8.25).

We have from eqn (8.24)

$$|c_{n+2}| \le M \left(\left| \sum_{k=0}^{n}\sum_{i=0}^{k} c_i c_{k-i} c_{n-k} \right| + |c_{n+1}| \right), \quad n = 1,\ 2,\ 3, \ldots,$$

where $M = \max\left\{ |\omega|, \frac{1}{\kappa^2} \right\}$.

Next, we define a power series

$$\mu = P(\xi) = \sum_{n=0}^{\infty} p_n \xi^n, \tag{8.26}$$

with $p_0 = |c_0|$, $p_1 = |c_1|, p_2 = |c_2|$ and $p_{n+2} = M\left(\left|\sum_{k=0}^{n}\sum_{i=0}^{k} c_i c_{k-i}\right.\right.$ $\left.c_{n-k}\right| + |c_{n+1}|\big), n = 1, 2, 3, \ldots,$ which implies that $|c_0| \leq p_n, n = 0, 1, 2, 3, \ldots.$

Thus, we may conclude that the series

$$\mu = P(\xi) = \sum_{n=0}^{\infty} p_n \xi^n, \tag{8.27}$$

is a majorant series of eqn (8.22).

Further we shall show that this series (8.27) has a positive radius of convergence.

Then, we have

$$\begin{aligned}
P(\xi) &= p_0 + p_1\xi^1 + p_2\xi^2 + \sum_{n=1}^{\infty} p_{n+2}\xi^{n+2} \\
&= p_0 + p_1\xi^1 + p_2\xi^2 \\
&\quad + M\left(\sum_{n=1}^{\infty} p_{n+2}\xi^{n+2} + \sum_{n=1}^{\infty}\sum_{k=0}^{n}\sum_{i=0}^{k} p_i p_{k-i} p_{n-k}\xi^{n+2}\right) \\
&= p_0 + p_1\xi^1 + p_2\xi^2 \\
&\quad + M\sum_{n=1}^{\infty}\left(p_{n+2}\xi^{n+2} + \sum_{k=0}^{n}\sum_{i=0}^{k} p_i p_{k-i} p_{n-k}\xi^{n+2}\right) \\
&= p_0 + p_1\xi^1 + p_2\xi^2 + M\left(-P^2(\xi) + p_0^2 + 2p_0p_1\xi + \left(p_1^2 + 2p_0p_2\right)\xi^2\right. \\
&\quad \left. + \left(2p_0p_3 + 2p_1p_2 + p_3\right)\xi^3\right).
\end{aligned} \tag{8.28}$$

Let us assume the implicit functional equation

$$\begin{aligned}
F(\xi, \mu) = \mu - p_0 + p_1\xi^1 + p_2\xi^2 + M\big(-\mu^2 + {p_0}^2 + 2p_0p_1\xi + \big({p_1}^2 \\
+ 2p_0p_2\big)\xi^2 + \left(2p_0p_3 + 2p_1p_2 + p_3\right)\xi^3\big) = 0.
\end{aligned}$$

As F is analytic in (ξ, μ)-plane, and $F(0, p_0) = 0$, $F'_{\mu}(0, p_0) = 1 + 2p_0M \neq 0$.

Therefore, by implicit function theorem [15, 16], we see that $\mu = P(\xi)$ is analytic in a neighborhood of point $(0, p_0)$ of the plane having a positive radius.

Hence, power series (8.28) is convergent in a neighborhood of the point $(0, p_0)$ of the plane.

8.5 Conclusion

The investigations of both untreated and cured brain tumors have been using mathematical models in recent years. In this research study, we have studied the brain tumor model, taking into consideration the well-known reaction–diffusion model with the help of Burgers' equation using the Lie symmetry method. Using Lie symmetry method, we have reduced PDEs to ODEs by determining infinitesimal generators and then applying the commutator bracket to find the commutation relation between infinitesimal generators. After that, we have found invariant solutions, by taking into new similarity variables with the help of obtained vector fields in each case. As the obtained ODEs are highly nonlinear systems that are of great relevance to real-world problems and the solution of such nonlinear systems is not an easy task, sometimes it may be possible that their exact solution may not be obtained. Exclusively, the solutions of the obtained ODEs are found with the help of the power series method, followed by discussing their convergence to establish the efficiency and efficacy of extracted solutions, which might have few significant implications in real-life situations.

Acknowledgments

All the authors express their sincere gratitude to *National Board for Higher Mathematics (NBHM) DAE, Govt. of India* for providing financial support under the research project, as evidenced by letter number: *02011/5/2023 NBHM (R. P)/ R&D-II/1140.*

References

[1] M. Wang, X. Li, J. Zhang, The $(G'/G)-$expansion method and travelling wave solutions of nonlinear evolution equations in mathematical physics, *Physics Letters A* (2008): 372 417-423.

[2] A. M. Wazwaz, A sine-cosine method for handling nonlinear wave equations, *Mathematical and Computational Modelling* (2004b): 40 499-508.

[3] S. Zhang, T. Xia, A generalized F-expansion method and new exact solutions of Konopelchenko-Dubrovsky equations, *Applied Mathematics and Computation* (2006): 183 1190-1200.

[4] A. M. Wazwaz, The tanh method for traveling wave solutions of nonlinear equations, *Applied Mathematics and Computation* (2004a): 154 713-723.

[5] E. Fan, Extended tanh-function method and its applications to nonlinear equations, *Physics Letters A* (2000): 277 212-218.

[6] E. M. E. Zayed, A. G. Al-Nowehy, The solitary wave ansatz method for finding the exact bright and dark soliton solutions of two nonlinear Schrödinger equations, *Journal of the Association of Arab Universities for Basic and Applied Sciences*, (2017): 24 184-190.

[7] P. Verma, L. Kaur, Extended $\exp(-\varphi\ (\xi))-$expansion method for generalized breaking soliton equation, *AIP Conference Proceedings*, (2020): 2214 020006.

[8] L. Kaur, A. M. Wazwaz, Optical soliton solutions of variable coefficient Biswas–Milovic (BM) model comprising Kerr law and damping effect, *Optik* (2022): 266 169617.

[9] S. Sarker, G. S. Said, M.M. Tharwat, R. Karim, M. Ali Akbar, N. S. Elazab, M.S. Osman, P. Dey, Soliton solutions to a wave equation using the (ϕ'/ϕ)– expansion method, *Partial Differential Equations in Applied Mathematics* (2023): 8 100587.

[10] G. W. Bluman, S. C. Anco, Symmetry and Integration Methods for Differential Equations, Springer, *Applied Mathematical Sciences* (2002).

[11] G. W. Bluman, J. D. Cole, The general similarity solution of the heat equation, *Journal of Mathematics and Mechanics* (1969): 18 1025-1042.

[12] J. D. Murray, Mathematical Biology II: Spatial Models and Biomedical Applications Third Edition, *Springer* (2003).

[13] S. Mayo, P. G. L. Leach, Symmetry Methods Applied to a Mathematical Model of a Tumor of the Brain, *Proceedings of Institute of Mathematics of NAS of Ukraine* (2004): 50 1 204-210.

[14] L. Kaur, R. K. Gupta, Kawahara equation and modified Kawahara equation with time dependent coefficients: symmetry analysis and generalized $(G'/G)-$expansion method, *Mathematical Methods in the Applied Sciences* (2012): 36 584-600.

[15] W. Rudin, Principles of Mathematical Analysis, *China Machine Press: Beijing* (2004).

[16] G. M. Fichtenholz, Functional Series, *Gordon and Breach: New York* (1970).

9

Convergence of Generalized Szász–Mirakjan–Kantorovich Operators on Different Function Spaces

Pooja Gupta and Ramu Dubey

Department of Mathematics, J. C. Bose University of Science and Technology, YMCA, India
E-mail: poojaguptaiitr@gmail.com; rdubeyjiya@gmail.com

Abstract

The aim of this study is to explore the behavior of a series of generalized operators known as the Szász–Mirakjan–Kantorovich operators within various function spaces, including the space of polynomials and the space containing real-valued continuous functions defined on $[0, \infty)$, which converge at infinity. These operators constitute a Kantorovich-style extension of the Jakimovski–Leviatan operators, based on multiple Appell polynomials. Our focus will be on analyzing the convergence rate of these operators using the moduli of smoothness.

Mathematical Subject Classification (2010): 41A36.

Keywords: Szász operator, multiple appell polynomials, moduli of smoothness

9.1 Introduction

A sequence of positive linear operators $(M_n)_{n\geq 1}$ were studied by G. M. Mirakjan [6], J. Favard [8], and O. Szász [13], which are defined as follows:

$$M_n(f)(x) = \sum_{m=0}^{\infty} e^{-nx} \frac{(nx)^k}{k!} f\left(\frac{k}{n}\right) \quad (n \geq 1, x \geq 0), \tag{9.1}$$

for all real-valued functions defined on positive axis for which the series at the right-hand side is absolutely convergent. The space of such functions includes, particularly, the space $\mathcal{A}([0,\infty))$ of all functions $f:[0,\infty)\to\mathbb{R}$ such that $|f(v)|\le M\exp(\beta v)(v\ge 0)$, for some $M\ge 0$ and $\beta\in\mathbb{R}$. Later on in order to approximate locally integrable functions on unbounded intervals, Butzer [11] defined an integral modification of the operators M_n for every $n\ge 1$ and $x\ge 0$, for the function space $\mathcal{B}([0,\infty))$, where $\mathcal{B}([0,\infty))$ is the space of all Borel measurable locally integrable functions $f:[0,\infty)\to\mathbb{R}$ such that the antiderivative of f, i.e., $F(x)=\int_0^x f(t)dt$ belongs to $\mathcal{A}([0,\infty))$ as follows:

$$L_n(f)(v)=n\sum_{m=0}^{\infty}e^{-nx}\frac{(nx)^k}{k!}\int_{\frac{k}{n}}^{\frac{k+1}{n}}f(t)dt.$$

These operators $(L_n)_{n\ge 1}$ were named as Szász–Mirakjan–Kantorovich operators. In more recent years, modifications of Szász–Mirakjan–Kantorovich operators including multiple Appell polynomials have been the object of investigation by several mathematicians. A system of multiple polynomials, referred to as the multiple polynomial system [5] and denoted by $\Im_{p_2,p_1}(x)$, is termed as multiple Appell if its generating function follows the structure depicted below:

$$\Re(y_1,y_2)e^{x(y_1+y_2)}=\sum_{p_2=0}^{\infty}\sum_{p_1=0}^{\infty}\frac{\Im_{p_2,p_1}(x)}{p_2!p_1!}y_1^{p_2}y_2^{p_1}, \tag{9.2}$$

where $\Re(y_1,y_2)$ has the series expansion given by

$$\Re(y_1,y_2)=\sum_{p_2=0}^{\infty}\sum_{p_1=0}^{\infty}\frac{a_{p_2,p_1}}{p_2!p_1!}y_1^{p_2}y_2^{p_1}, \tag{9.3}$$

with $\Re(0,0)=a_{0,0}\neq 0$ and $a_{p_2,p_1}\ge 0$ for all $p_2,p_1\in\mathbb{N}$. Also, eqn (9.2) and (9.3) converge for $|y_1|\le S_1$, $|y_2|\le S_2$ $(S_1,S_2>1)$. The multiple polynomial system $\Im_{p_2,p_1}$ satisfies the relationship:

$$\Im'_{p_2,p_1}(x)=p_2\Im_{p_2-1,p_1}(x)+p_1\Im_{p_2,p_1-1}(x),$$

for every $p_2+p_1\ge 1$. Furthermore, for these multiple Appell polynomial systems $\Im_{p_2,p_1}(x)$, there exists a unique sequence $\{a_{p_2,p_1}\}_{p_2,p_1=0}^{\infty}$ with $a_{0,0}\neq 0$, such that:

$$\Im_{p_2,p_1}(x)=\sum_{d_1=0}^{p_2}\sum_{d_2=0}^{p_1}\binom{p_2}{d_1}\binom{p_1}{d_2}a_{p_2-d_1,p_1-d_2}x^{d_1+d_2}. \tag{9.4}$$

Thus, $\Im_{p_1,p_2}(x)$ is a polynomial in x with a degree of p_1+p_2. Varma [19] extended the Szász operators [13] using these multiple Appell polynomials as follows:

$$v_n(f;v) = \frac{e^{-nv}}{\Re(1,1)} \sum_{p_2=0}^{\infty} \sum_{p_1=0}^{\infty} \frac{\Im_{p_2,p_1}\left(\frac{nv}{2}\right)}{p_2!p_1!} f\left(\frac{p_2+p_1}{n}\right). \tag{9.5}$$

He also introduced the Kantorovich-type modification of these operators as

$$K_n^*(r;v) = \frac{ne^{-nv}}{\Re(1,1)} \sum_{p_2=0}^{\infty} \sum_{p_1=0}^{\infty} \frac{\Im_{p_2,p_1}\left(\frac{nv}{2}\right)}{p_2!p_1!} \int_{\frac{p_2+p_1}{n}}^{\frac{p_2+p_1+1}{n}} r(w)dw,$$

and analyzed the convergence rate of these operators in terms of the classical modulus of continuity.

The objective of this chapter is to explore the generalization of these operators, defined by Swarup et al. [4], as

$$\varphi_n(q;v) = \frac{ne^{-nv}}{(f_n-g_n)\Re(1,1)} \sum_{p_2=0}^{\infty} \sum_{p_1=0}^{\infty} \frac{\Im_{p_2,p_1}\left(\frac{nv}{2}\right)}{p_2!p_1!} \int_{\frac{p_2+p_1+g_n}{n}}^{\frac{p_2+p_1+f_n}{n}} q(w)dw \quad (n \geq 1,\ v \geq 0). \tag{9.6}$$

For every $q \in \Gamma[0,\infty)$ (the space of all Borel measurable and locally integrable functions $c(v) : [0,\infty)$ to $\mathbb{R}$ such that the antiderivative $C(v) = \int_0^v c(t)dt (v \geq 0)$ belongs to $\phi([0,\infty))$: the space of all functions $s : [0,\infty)$ to $\mathbb{R}$ such that $|s(v)| \leq Ue^{hv} (v \geq 0)$ for some $U \geq 0$ and $h \in \mathbb{R}$). Here $(g_n)_{(n\geq 1)}$ and $(f_n)_{(n\geq 1)}$ continue as given.

The chapter investigates the approximation properties of (φ_n) for every $n \geq 1$ on various continuous and weighted continuous function spaces and examines the convergence rate of these operators using suitable moduli of smoothness. Relevant literature related to this study includes [14–16].

9.2 Literature Review

To establish the renowned Weierstrass' [10] theorem, which asserts that any continuous function defined on a closed interval $[a,b]$ can be uniformly approximated with arbitrary precision by a polynomial function, Bernstein presented a straightforward proof employing Bernstein polynomials, which are associated with probability theory. He demonstrated that for a continuous function $f(x)$ defined on the unit interval, and any given $\epsilon > 0$, it is possible to find a polynomial $P_n(x) = b_0x^n + b_1x^{n-1} + \cdots + b_n$ of degree n, satisfying the inequality:

$$|f(x) - P_n(x)| < \epsilon, \tag{9.7}$$

for every point x in the unit interval. Bernstein achieved this by considering an event B with probability x and utilizing it as follows:

Imagine conducting n experiments and rewarding a player with the sum $f\left(\frac{m}{n}\right)$ if event B occurs m times. Under these conditions, the player's expected value can be expressed as

$$E_n(x) = \sum_{m=0}^{n} f\left(\frac{m}{n}\right)\binom{n}{m}x^m(1-x)^{n-m}.$$

Bernstein [18] demonstrated that these polynomials E_n precisely satisfy eqn (9.7). Consequently, these polynomials serve as effective approximations of f across the unit interval. Subsequently, a generalization was devised to approximate f over the interval $[a, b]$. Bernstein polynomials find numerous applications in computer graphics for curve and surface modeling, serving as the foundation for Bezier curves and surfaces. These elements are extensively utilized in computer-aided design (CAD) and computer-aided manufacturing (CAM) for shaping and curve modeling purposes.

Moreover, Bernstein polynomials have applications in probability theory and statistics, particularly in non-parametric estimation techniques like kernel density estimation and non-parametric regression. They offer a versatile framework for estimating probability density functions and regression functions from empirical data. Additionally, Bernstein polynomials contribute to control theory for system identification and control system design. They play a role in polynomial approximation-based control algorithms such as polynomial chaos expansion (PCE) methods, aiding in modeling uncertainties in dynamic systems and designing robust control strategies.

In subsequent developments, Lupas devised a method to approximate Lebesgue integrable functions f over the interval $[0, 1]$ using a sequence of beta operators. These operators, defined as

$$B_n(f, x) = \int_0^1 \beta_n(v, x)f(v)dv,$$

involve a beta distribution function $\beta_n(v, x)$ given by

$$\frac{1}{B(nx+1, n+1-nx)} u^{nx}(1-u)^{n-nx},$$

where $B(p, q)$ represents the beta function.

Subsequently, various other sequences and classes of linear positive operators emerged, such as Kantorovich operators [20] and Durrmeyer operators [9]. These operators leverage integral modifications of Bernstein polynomials to effectively approximate Lebesgue integrable functions defined on the interval $[0, 1]$.

Kantorovich operators are pivotal in signal approximation and noise reduction tasks. They are integral in signal denoising, interpolation, and spectral analysis, where precise signal approximation is critical for subsequent processing and analysis. Additionally, these operators are employed in various image processing applications, including image restoration, enhancement, and compression. By employing Kantorovich operators to approximate images, it becomes feasible to diminish noise, enhance image quality, and compress images efficiently for storage and transmission purposes.

Further, utilizing the Poisson distribution, Szász introduced a sequence of operators, as defined in eqn (9.1), aimed at approximating continuous functions defined on the positive real axis.

In 1969, Jakimovski and Leviatan [2] generalized Szász operators by incorporating Appell polynomials, establishing various approximation properties for real and analytic functions along with their derivatives. Appell polynomials, named after Paul Émile Appell, hold significant importance in mathematical analysis, particularly within the realm of hypergeometric functions. These polynomials are a class of special functions that find extensive application in various fields. They serve as essential tools for solving certain differential equations and recurrence relations. Moreover, their significance extends beyond mathematical analysis to include mathematical physics and theoretical mathematics.

Subsequently, in 2017, Mursaleen [12] introduced Stancu-type Jakimovski–Leviatan operators and investigated their approximation properties. In 2011, Lee [5] defined multiple Appell polynomials and derived several equivalent conditions for this polynomial class. Utilizing these multiple Appell polynomials, Varma [19] extended the Szász operators and explored various approximation properties for continuous functions defined on the positive real axis.

The applications of multiple Appell polynomials encompass a wide range of scientific and engineering fields, facilitating the resolution of diverse problems entailing multidimensional phenomena, nonlinear dynamics, and intricate interactions. Their utility continues to expand with the development of novel mathematical methodologies and computational techniques.

9.3 Generalizing Jakimovski–Leviatan Operators

For a given $c \in \Gamma([0,\infty))$ and $C(v) = \int_0^v c(w)dw$, the antiderivative of $c(v)$ operators given in eqn (9.6) may be expressed as

$$\begin{aligned}\varphi_n(c;v) &= \frac{ne^{-nv}}{(f_n - g_n)\Re(1,1)} \sum_{p_2=0}^{\infty}\sum_{p_1=0}^{\infty} \frac{\Im_{p_2,p_1}(\frac{nv}{2})}{p_2!p_1!} \\ &\quad \left[C\left(\frac{p_2+p_1+f_n}{n}\right) - C\left(\frac{p_2+p_1+g_n}{n}\right)\right], \\ &= \frac{n}{(f_n - c_n)} \upsilon_n(\Gamma_n(C))(v),\end{aligned}$$

where υ is given by eqn (9.5) and the mapping Γ_n is defined as

$$\Gamma_n(C)(v) := C\left(v + \frac{f_n}{n}\right) - C\left(v + \frac{g_n}{n}\right), \quad (v \geq 0).$$

In this chapter, we will consistently use the following notations.

For any number m that is greater than or equal to 1, we define a function $v_m(x)$ as $(1+x^m)^{-1}(x \geq 0)$ and, additionally, we denote a set $S_m(x)$ as follows:

$$S_m := \left\{f \in \Gamma([0,\infty)) sup_{x\geq 0} v_m(x)|f(x)| \in \mathbb{R}\right\},$$

where $\Gamma([0,\infty))$ denotes the space of all real-valued continuous functions on $[0,\infty)$.

S_m forms a Banach lattice when equipped with both point-wise ordering and a weighted norm.

$$||f||_m = \sup_{x\geq 0} v_m(x)|f(x)|.$$

Next, we define the following spaces:

$$S_m^* = \left\{f \in S_m| \lim_{x\to\infty} v_m(x)f(x) \in \mathbb{R}\right\}$$

and

$$S_m^0 = \left\{ f \in S_m^* | \lim_{x \to \infty} v_m(x) f(x) = 0 \right\}.$$

That leads to the formation of Banach sublattices of S_m.

9.4 Preliminaries

Lemma 9.1. For operators (9.6), the estimates of moments are as follows [19]:

(i) $\varphi_n(1; v) = 1;$

(ii) $\varphi_n(w; v) = v + \frac{1}{2} \frac{f_n + g_n}{n} + \frac{\Re_{y_1}(1,1) + \Re_{y_2}(1,1)}{n\Re(1,1)};$

(iii) $\varphi_n(w^2; v) = v^2 + \frac{v}{n}\left(1 + \frac{f_n + g_n}{n} + 2\frac{(\Re_{y_1}(1,1) + \Re_{y_2}(1,1))}{\Re(1,1)}\right) + \frac{f_n^2 + g_n^2 + f_n g_n}{3n^2} \times \frac{1}{n^2\Re(1,1)}\{\Re_{y_1}(1,1)(1 + f_n + g_n) + \Re_{y_2}(1,1)(1 + f_n + g_n) + \Re_{y_1y_1}(1,1) + \Re_{y_2y_2}(1,1) + 2\Re_{y_1y_2}(1,1)\};$

(iv) $\varphi_n(\xi_v(w); v) = \frac{1}{2} \frac{f_n + g_n}{n} + \frac{\Re_{y_1}(1,1) + \Re_{y_2}(1,1)}{n\Re(1,1)};$

(v) $\varphi_n(\xi_v^2(w); v) = \frac{v}{n} + \frac{1}{n^2\Re(1,1)}\{\Re_{y_1}(1,1) + \Re_{y_2}(1,1) + \Re_{y_1y_1}(1,1) + \Re_{y_2y_2}(1,1) + 2\Re_{y_1y_2}(1,1)\} + \frac{g_n^2 + f_n^2 + g_n f_n}{n^2} + \frac{1}{n}(f_n + g_n)\left(\frac{\Re_{y_1}(1,1) + \Re_{y_2}(1,1)}{n\Re(1,1)}\right).$

For every $\mu > 0$, let $m_\mu(v) = e^{-\mu v}$, where $v \geq 0$. Then from ([4], Proposition 1),

$$v(m_\mu)(v) = \frac{\Re(e^{\frac{-\mu}{n}}, e^{\frac{-\mu}{n}})}{\Re(1,1)} exp\left(nv\left(e^{\frac{-\mu}{n}} - 1\right)\right) \tag{9.8}$$

and

$$\varphi_n(m_\mu)(v) = \frac{n}{\mu(f_n - g_n)}\left(e^{-\frac{\mu g_n}{n}} - e^{-\frac{\mu f_n}{n}}\right) v(m_\mu). \tag{9.9}$$

Value of the generalized Szász operators defined in eqn (9.5) for the functions $\phi_m(x) = x^m$ for every $n \geq 1$ and $m \geq 0$ is as follows [4]:

$$\upsilon_n(t^m; x) = \sum_{j=0}^{m} b_{m,m-j}(1,1)n^{j-m}\phi_j, \tag{9.10}$$

where $b_{m,0} = 1$ and $b_{m,i}(t_1, t_2)$ is a function of t_1 and t_2, and this function is nothing but a linear combination of the partial derivatives of function $\Re(t_1, t_2)$ and the order of the partial derivative is maximum m for each $i \in \{1, 2, ..., m-1, m\}$.

For every $n \geq 1$ and $m \geq 0$,

$$\begin{aligned}
\varphi_n(\phi_m) &= \frac{e^{-nv}}{(m+1)(f_n - g_n)\Re(1,1)n^m} \sum_{p_2=0}^{\infty}\sum_{p_1=0}^{\infty} \frac{\Im_{p_2,p_1}(\frac{nv}{2})}{p_2!p_1!} \\
&\quad \left((p_2 + p_1 + f_n)^{m+1} - (p_2 + p_1 + g_n)^{m+1}\right), \\
&= \frac{e^{-nv}}{(m+1)(f_n - g_n)\Re(1,1)n^m} \sum_{p_2=0}^{\infty}\sum_{p_1=0}^{\infty} \frac{\Im_{p_2,p_1}(\frac{nv}{2})}{p_2!p_1!} \left[\binom{m+1}{0}(f_n^{m+1} - g_n^{m+1})\right. \\
&\quad + \binom{m+1}{1}(p_1+p_2)(f_n^m - g_n^m) + \binom{m+1}{2}(p_1+p_2)^2(f_n^{m-1} - g_n^{m-1}) \\
&\quad \left. + \cdots + \binom{m+1}{m}(p_1+p_2)^m(f_n - g_n)\right] \\
&= \frac{e^{-nv}}{(m+1)\Re(1,1)n^m} \sum_{p_2=0}^{\infty}\sum_{p_1=0}^{\infty} \frac{\Im_{p_2,p_1}(\frac{nv}{2})}{p_2!p_1!} \\
&\quad \left[\sum_{k=0}^{m}\binom{m+1}{k}\left(\sum_{p=0}^{m-k} f_n^p g_n^{m-k-p}\right)(p_1+p_2)^k\right] \\
&= \frac{1}{(m+1)n^m}\left[\sum_{k=0}^{m}\binom{m+1}{k}\left(\sum_{p=0}^{m-k} f_n^p g_n^{m-k-p}\right)\right. \\
&\quad \left.\sum_{p_2=0}^{\infty}\sum_{p_1=0}^{\infty} \frac{e^{-nv}\Im_{p_2,p_1}(\frac{nv}{2})}{\Re(1,1)p_2!p_1!}(p_1+p_2)^k\right] \\
&= \frac{1}{(m+1)n^m}\sum_{k=0}^{m}\binom{m+1}{k}\left(\sum_{p=0}^{m-k} f_n^p g_n^{m-k-p}\right)\upsilon_n(t^k)n^k \\
&= \frac{1}{(m+1)n^m}\sum_{k=0}^{m}\binom{m+1}{k}\left(\sum_{p=0}^{m-k} f_n^p g_n^{m-k-p}\right)\sum_{j=0}^{k} b_{k,k-j}(1,1)n^j\phi_j \\
&= \phi_m + \frac{1}{n}Q_{m-1},
\end{aligned}$$

where the coefficients $b_{k,k-j}(1,1)$ are the same as given in eqn (9.10) and Q_{m-1} is a positive polynomial with degree $m-1$. From this, we can conclude that $\phi_m \leq \varphi_n(\phi_m)$ for each $m \geq 0$.

Further, let $\mathbb{R}_m$ denote the space of all polynomials of degree $m, m \geq 1$, then

$$\varphi_n(\mathbb{R}_m) \subset \mathbb{R}_m,$$

for each $m \geq 1$.

Further, for every $m \geq 1, n \geq 1$ and $x \geq 0$,

$$v_m(x)\varphi_n(\phi_m)(x) \leq v_m(x)\phi_m(x) + \frac{k_m}{n}, \tag{9.11}$$

where

$$K_m := \max_{x \geq 0} v_m(x) \left\{ \sum_{j=0}^{m-1} b_{m,m-j} x^j + \sum_{j=0}^{m-1} \binom{m}{k} \sum_{j=0}^{k} b_{k,k-j} x^j \right\}, \tag{9.12}$$

and $v_m(x) = (1 + x^m)^{-1} (x \geq 0)$.

Therefore, for each $m \geq 0$,

$$\lim_{n \to \infty} ||\varphi_n(\phi_m) - \phi_m||\,|_m = 0. \tag{9.13}$$

Finally, for every $n, m \geq 1$ and $x \geq 0$,

$$\varphi_n(\rho_x^m) = \sum_{a=0}^{m} \binom{m}{a} \frac{(-1)^{m-a} x^{m-a}}{(a+1)n^a}$$

$$\sum_{k=0}^{a} \binom{a+1}{k} \sum_{p=0}^{a-k} f_n^p g_n^{a-k-p} \sum_{j=0}^{k} b_{k,k-j} n^j \phi_j, \tag{9.14}$$

where $\rho_x(y) = y - x (y \geq 0)$.

9.5 Approximation Properties

Theorem 9.1. Consider the positive operators $\varphi_n, n \geq 1$, defined by eqn (9.6). Then for $n \geq 1$ and $m \geq 1$ fixed, *(a)* φ_n is a positive continuous linear operator from S_m into itself and $||\varphi_n||_{S_m} \leq 1 + \frac{K_m}{n}$, K_m is as defined in eqn (9.12); in particular,

$$\sup_{n\geq 1} ||\varphi_n||_{S_m} \leq 1 + K_m; \tag{9.15}$$

(b) $\varphi_n(S_m^0) \subset S_m^0$.

Proof. From Proposition 9.4 and eqn (9.11), we have

$$||\varphi_n||_{S_m} \subset S_m$$

and

$$v_m(x)|\varphi_m(f)(x)| \leq ||f||_m v_m(x)\varphi_n(1 + v_m)(x) \leq ||f||_m \left(1 + \frac{k_m}{n}\right).$$

For part (b), consider the subspace B generated by the family of functions $(m_\mu)_{\mu\geq 0}$ defined as $m_\mu(x) = e^{-\mu x}$, which by the Stone–Weierstrass theorem is dense in $\xi_0([0,\infty))$ and hence in S_m^0. Since

$$\varphi_n(B) \subset \xi_0([0,\infty)) \subset S_m^0,$$

part (b) is proved. □

Theorem 9.2. If for a given $m \geq 1,\ \ f \in S_m^*$, then

$$\lim_{n\to\infty} v_m(\varphi_n(f) - f) = 0 \qquad uniformly\ on\ \ [0,\infty) \tag{9.16}$$

i.e., $\lim_{n\to\infty} \varphi_n(f) = f$ with respect to $||.||_m$.

If $f \in S_m$, then

$$\lim_{n\to\infty} v_m(\varphi_n(f) - f) = 0 \tag{9.17}$$

uniformly on compact subset of $[0,\infty)$.

Proof. Consider the functions $m_\mu(t) = e^{-\mu t}$, $\mu > 0$. Then $\lim_{n\to\infty} \varphi_n(m_\mu) = m_\mu$ with respect to $||.||_\infty$ and therefore with respect to $||.||_m$. Since the sequence $(\varphi_n)_{n\geq 1}$ is equibounded on S_m^0 and the linear subspace generated by $(m_\mu)_{\mu\geq 0}$ is dense in S_m^0, eqn (9.16) is true for $f \in S_m^0$. On the other hand, if $f \in S_m^*$, then $f = h + \gamma_m(1 + \phi_m)$, where $\gamma_m = \lim_{x\to\infty} \phi_m(x)f(x) \in \mathbb{R}$

and $g = f - \gamma_m(1 + \phi_m) \in S^0_m$. Therefore, eqn (9.16) follows for f too on account of eqn (9.13).

The preceding result and the inclusion $S_m \subset S^0_{m+1}$ gives eqn (9.17), because if W is any compact subset of $[0, \infty)$, then

$$v_m(x)|\varphi_n(f)(x) - f(x)| \leq K\,||\varphi_n(f) - f|\,|_m \tag{9.18}$$

for each $x \in W$, where $K = \sup\limits_{x \in W} \dfrac{v_m(x)}{v_{m+1}(x)}$. □

Theorem 9.3. Let $(\varphi_n)_{n\geq 1}$ be the sequence of operators defined by eqn (9.6). Then,

$$|\varphi_n(m_\mu)(-\log x) - x^\mu| \leq \frac{5\mu}{4n} + \frac{c}{n^2}, \tag{9.19}$$

for every $\mu > 0, n \geq 1$ and $0 < x \leq 1$, where $m_\mu(t) = e^{-\mu}(t)$ and c is some constant depending on μ.

Proof. Since for every $0 \leq x \leq 1$, there exists an $j > 0$ such that $x = e^{-j}$. Therefore, from [[1], Lemma 3.1],

$$x^{n\left(1-e^{-\frac{\mu}{n}}\right)} - x^\mu \leq \frac{\mu}{2ne},$$

and by using eqn (9.3)

$$\begin{aligned}
\Re(e^{-\frac{\mu}{n}}, e^{-\frac{\mu}{n}}) &= \sum_{p_2=0}^{\infty}\sum_{p_1=0}^{\infty} \frac{a_{p_2,p_1}}{p_2!p_1!} e^{-\frac{\mu p_2}{n}} e^{-\frac{\mu p_1}{n}} \\
&= \sum_{p_2=0}^{\infty}\sum_{p_1=0}^{\infty} \frac{a_{p_2,p_1}}{p_2!p_1!} e^{-\frac{\mu(p_1+p_2)}{n}} \\
&\leq \sum_{p_2=0}^{\infty}\sum_{p_1=0}^{\infty} \frac{a_{p_2,p_1}}{p_2!p_1!} \left(1 + \frac{c}{n^2}\right),
\end{aligned}$$

where c is some constant depending on μ.

Therefore,

$$\frac{\Re(e^{-\frac{\mu}{n}}, e^{-\frac{\mu}{n}})}{\Re(1,1)} = \left(1 + \frac{c}{n^2}\right).$$

Then by using eqn (9.8) and (9.9), and [[3], equation (3.2)], for every $n \geq 1, \mu > 0$ and $0 < x \leq 1$, we have

$$
\begin{aligned}
&|\varphi_n(m_\mu)(-\log x) - x^\mu| \\
&= \left| \frac{n}{\mu(f_n - g_n)} \frac{\Re(e^{-\frac{\mu}{n}}, e^{-\frac{\mu}{n}})}{\Re(1,1)} \left(e^{-\frac{\mu g_n}{n}} - e^{-\frac{\mu f_n}{n}}\right) x^{n(1-e^{-\frac{\mu}{n}})} - x^\mu \right| \\
&= \frac{n}{\mu(f_n - g_n)} \frac{\Re(e^{-\frac{\mu}{n}}, e^{-\frac{\mu}{n}})}{\Re(1,1)} \left(e^{-\frac{\mu g_n}{n}} - e^{-\frac{\mu f_n}{n}}\right) \left| x^{n(1-e^{-\frac{\mu}{n}})} - x^\mu \right| \\
&\quad + x^\mu \frac{\Re(e^{-\frac{\mu}{n}}, e^{-\frac{\mu}{n}})}{\Re(1,1)} \left(1 - \frac{n}{\mu(f_n - g_n)} \left(e^{-\frac{\mu g_n}{n}} - e^{-\frac{\mu f_n}{n}}\right)\right) \\
&\quad + x^\mu \left| \frac{\Re(e^{-\frac{\mu}{n}}, e^{-\frac{\mu}{n}})}{\Re(1,1)} - 1 \right| \\
&\leq \frac{n}{\mu(f_n - g_n)} e^{-\frac{\mu g_n}{n}} \left(1 - e^{-\frac{\mu(f_n - g_n)}{n}}\right) \frac{\mu}{2ne} + \frac{\mu}{n} + \frac{c}{n^2} \\
&\leq \frac{5\mu}{4n} + \frac{c}{n^2}.
\end{aligned}
$$

□

Next, to find some uniform estimates, we will use a similarity technique, i.e., given an approximation process $(R_n)_{n\geq 1}$ on some Banach space R, if $\Omega : R \to X$ is an isometric between R and another Banach space X, then we can construct an approximation process on X by taking

$$R_n^* = \Omega \circ R_n \circ \omega^{-1} \quad (n \geq 1).$$

In this case, $(R_n)_{n\geq 1}$ and (R_n^*) are said to be isomorphic and for each $r \in R$,

$$||R_n(r) - r||_R = ||R_n^*(\Omega(r)) - \Omega(r)||_X, \tag{9.20}$$

which means that instead of finding the rate of convergence of R_n, the rate of convergence of the sequence R_n^* can be found.

As an example of this isomorphism, first take the isomorphism $\Omega : \varrho_*([0,\infty)) \to \varrho([0,1])$ defined as

$$\Omega(r)(v) = \begin{cases} r(-\log t) & \text{if } 0 < v \leq 1; \\ \lim\limits_{x\to\infty} r(x) & \text{if } v = 0 \end{cases}$$

for each $f \in \varrho_*([0,\infty))$, where $\varrho_*([0,\infty))$ denotes the space of all continuous functions defined on $[0,\infty)$, which converges at infinity and $\varrho([0,1])$ is the space of all continuous functions defined on the unit interval.

Here the inverse of the function Ω, i.e., $\Omega^{-1} : \varrho([0,1]) \to \varrho_*([0,\infty))$ is defined as

$$\Omega^{-1}(h)(t) := h(e^{-t})$$

for every $h \in \varrho([0,1])$ and $t \geq 0$.

Moreover, for every $n \geq 1$ and $h \in \varrho([0,1]$, we take

$$\varphi_n * (h) := \Omega(\varphi_n(\Omega^{-1}(h))). \tag{9.21}$$

It can be seen easily that $\varphi_n^*(\mathbf{1}) = \mathbf{1}$,

$$\varphi_n^*(\xi_v)(u) = \begin{cases} \varphi_n(m_1 - v\mathbf{1})(-\log u) & \text{if } 0 < u \leq 1; \\ 0 & \text{if } u = 0 \end{cases} \tag{9.22}$$

and

$$\varphi_n^*(\xi_v)^2(u) = \begin{cases} \varphi_n(m_2 - 2vm_1 + v^2\mathbf{1})(-\log u) & \text{if } 0 < u \leq 1; \\ 0 & \text{if } u = 0. \end{cases} \tag{9.23}$$

$(v \in [0,1])$, where $m_\mu = e^{-\lambda\mu}$ for $\mu = 1, 2$ and $\xi_v(w) = w - v (w \geq 0)$.

Theorem 9.4. Let $(\varphi_n)_{n\geq 1}$ be the sequence of linear and positive operators defined by eqn (9.6) and acting on the space $\varrho_*([0,\infty))$. Then, for every $n \geq 1$ and $r \in \varrho_*([0,\infty))$,

$$\begin{aligned} ||\varphi_n(r) - r||_\infty \leq & \left(\frac{5}{4\sqrt{n}} + \frac{c}{n^{3/2}}\right)\omega\left(\Omega(r), \frac{1}{\sqrt{n}}\right) \\ & + \left(\frac{7}{2} + \frac{3c}{2n}\right)\omega_2\left(\Omega(r), \frac{1}{\sqrt{n}}\right). \end{aligned}$$

Proof.

Using the inequality denoted as eqn (9.21), we derive a uniform estimate for $||\varphi_n^*(\Omega(r)) - \Omega(r)||\infty$. To accomplish this, we employ Theorem (2.2.1) from the reference [17] and also refer to Theorem 10 in [7]. This applies for every $n \geq 1$, $r \in \varrho([0,\infty))$ where $0 \leq v \leq 1$, and $\delta > 0$,

$$\begin{aligned} |\varphi_n^*(\Omega(r))(v) - \Omega(r)(v)| \leq & |\varphi_n^*(\mathbf{1}) - 1|\Omega(r))(v)| + \frac{1}{\delta}|\varphi_n^*(\xi_v)(v)|\omega(\Omega(r), \delta) \\ & + \left(\varphi_n^*(\mathbf{1})(v) + \frac{1}{2\delta^2}\varphi_n^*(\xi_v^2)(v)\right)\omega_2(\Omega(r), \delta). \end{aligned} \tag{9.24}$$

From eqn (9.19) and (9.22) it follows that

$$|\varphi_n^*(\xi_v)(v)| \leq \frac{5}{4n} + \frac{c}{n^2} \tag{9.25}$$

and by (9.23), we have

$$\begin{aligned}\varphi_n^*(\xi_v^2)(v) &= \varphi_n(m_2)(-\log v) - v^2 - 2v(\varphi_n(m_1)(-\log v) - v)\\ &\leq \frac{5}{2n} + \frac{c}{n^2} + \frac{5}{2n} + 2\frac{c}{n^2} = \frac{5}{n} + 3\frac{c}{n^2}.\end{aligned} \tag{9.26}$$

□

9.6 Conclusion

In conclusion, this study explores the extensive applications and properties of mathematical operators and polynomials, including Jakimovski–Leviatan operators, and Appell polynomials. These tools offer powerful means for approximating functions across various domains, from signal and image processing to probability theory and control systems. The analysis of their properties, such as convergence and approximation accuracy, provides valuable insights into their utility and limitations. Overall, the study underscores the significance of these mathematical techniques in addressing complex problems in scientific and engineering disciplines, paving the way for further advancements in computational methodologies and theoretical understanding.

Acknowledgments

The authors are highly grateful to the Department of Science and Technology, India for the DST-PURSE grant (SR/PURSE/2022/126) financial support.

References

[1] Holhoş A. The rate of approximation of functions in an infinite interval by positive linear operators. *Studia Universitatis Babeş–Bolyai Math*, 55(2):133–142, 2010.

[2] Jakimovski A. and Leviatan D. Generalized Szász operators for the approximation in the infinite interval. *Mathematica (Cluj)*, 11:97–103, 1969.

[3] Montano C.M. Altomare F. and Leonessa V. On a generalization of Szász–Mirakjan–Kantorovich operators. *Results in Mathematics*, 63:837–863, 2013.

[4] Swarup C., Gupta P., Dubey R., and Mishra V. N. Generalization of Szász–Mirakjan–Kantorovich operators using multiple Appell polynomials. *Journal of Inequality and Applications*, 2020:1–11, 2020.

[5] Lee D.W. On multiple Appell polynomials. *Proceedings of the American Mathematical Society*, 139:2133–2141, 2011.

[6] Mirakjan G.M. Approximation of continuous functions with the aid of polynomials(russion), *Dokl. Akad. Naauk SSSR*, 31:201–205, 1941.

[7] Gonska H. Positive operators and approximation of functions: selected topics. *Conf. Semin. Mat. Univ. Bari, 288*, 28, 2002.

[8] Favard J. Sur les multiplicateurs d'interpolations. *Journal de Mathématiques Pures et Appliquées*, 23(9):219–247, 1944.

[9] Durrmeyer J.L. Une formule d'inversion de la transformée de laplace: Applications á la théorie des moments. *PhD thesis, These de 3e cycle*, 1967.

[10] Anatoliĭ Aleksandrovich Kilbas, Hari M Srivastava, and Juan J Trujillo. Uber die analytische darstellbarkeit sogenannter willkurlicher functionen einer reellen veranderlichen sitzungsberichteder. *Koniglich Preussischen Akademie der Wissenschcaften zu Berlin*, 204, 1885.

[11] Butzer P. L. On the extensions of bernstein polynomials to the infinite interval. *Proceedings of the American Mathematical Society*, 5(4):547–553, 1954.

[12] Mursaleen M. and Khan T. Approximation of continuous functions with the aid of polynomials. *Azerbaijan Journal of Mathematics*, 7(1):16–26, 2017.

[13] Szász O. Generalization of Bernstein's polynomials to the infinite interval. *J Res Nat Bur Stds*, 45:239–245, 1950.

[14] Gupta P., Ram M., and Dubey R. Image of polynomials under generalized Szász operators. *Applied Mathematics-A Journal of Chinese Universities*, 36(4):599–610, 2021.

[15] Gupta P. and Agrawal P.N. Jakimovski-Leviatan operators of Durrmeyer type involving Appell polynomials. *Turkish Journal of Mathematics*, 42:1457–1470, 2018.

[16] Gupta P. and Agrawal P.N. Quantitative Voronovskaja and Grüss Voronovskaja-type theorems for operators of Kantorovich type involving multiple Appell polynomials. *Iranian Journal of Science and Technology Transaction A, Science*, 43, 2019.

[17] Păltănea R. *Approximation theory using positive linear operators.* Birkhäuser, 2004.

[18] Bernstein S. Démonstration du théoréme de weierstrass fondée sur le calcul des probabilités. *Kharkiv Mathematical Society*, 13(1):1–2, 1912.

[19] Verma S. On a generalization of Szász operators by multiple Appell polynomials. *Studia Universitatis Babeş–Bolyai Math*, 58(3):361–369, 2013.

[20] Kantorovich L. V. Sur certains developments suivant les polynomes de la forme de s. Bernstein i, ii. *Dokl. Akad. Nauk SSSR*, 1-2:563–568, 595–600, 1930.

10

Herd Behavior in Prey and Cooperative Hunting in Predators With Holling Type-II: A Dynamical Approach

Kuldeep Singh[1], Ramu Dubey[1], and Teekam Singh[2]

[1]Department of Mathematics, J.C. Bose University of Science and Technology, YMCA, India
[2]Department of Computer Science and Engineering, Graphic Era (Deemed to be) University, India
E-mail: kuldeepsinghbhati599@gmail.com; rdubeyjiya@gmail.com; teekamsingh.iitr@gmail.com

Abstract

The development of functional responses to explain prey's group defense has been the topic of a recent study. This research investigates a predator–prey model that combines cooperative predator conduct with herd behavior in prey, along with Holling type-II functional responses in the prey–predator system. Our primary goal in this essay is to examine and study the temporal and spatial dynamics of the prey–predator system. The consequences of cooperative hunting and the spatial patterns arising from the fluctuation of the diffusion coefficient and the rate of cooperation hunting are of special interest to us. We confirm the boundedness and positivity of the solution of the proposed model. Next, the stability of the equilibrium point where populations of prey and predators coexist is examined. We talk about Hopf bifurcations that occur close to the equilibrium points. A stability investigation of these amplitude equations reveals several Turing patterns, including mixed, stripes, and spots. Next, we demonstrate the analytical findings with numerical simulations.

Keywords: Prey–predator system, Holling type-II functional, cooperative Hunting, herd behavior

10.1 Introduction

The complex web of relationships that exists in nature has long enthralled scientists. One of the most fascinating dynamics is that which exists between predators and prey. Studying these interactions offers important insights into the mechanisms controlling species survival and evolution in addition to illuminating the delicate balance of ecosystems. In order to comprehend the dynamics of these systems, academics have used a variety of mathematical models over the years, each of which has added a new level of detail.

Here, we explore the intriguing world of prey–predator systems in an effort to gain a better understanding by fusing two crucial components: cooperative hunting in predators and herd behavior in prey, along with the Holling type-II functional response. This unique model considers the predation rate as a function of prey density, accounting for the nuances of both predator cooperation and prey aggregation behavior. Prey animals frequently exhibit herd behavior, which is a key factor in determining how they formulate their survival tactics. Prey frequently congregate in groups, which provides safety through heightened awareness, reduced personal danger, and enhanced predator detection. However, a number of species have been shown to engage in cooperative hunting, which allows them to take down larger or more evasive prey. We seek to understand the emergent features and synergistic effects that result from the confluence of these two events by investigating their interaction.

We also include the Holling type-II functional response, which is a popular mathematical paradigm in the research on predator–prey relationships. The link between the predator's rate of intake and the available prey density is described by this functional response. It is important to comprehend this relationship because it accounts for variables like handling time and prey encounter frequencies, capturing the nonlinear nature of predation. Through the integration of these components, our research aims to offer a thorough and intricate examination of the interactions between prey and predator. Our research could provide fresh perspectives on the fundamental processes regulating the stability of ecosystems, the coexistence of predators and prey, and the possible effects of outside influences like climatic shifts. Furthermore, this research has applications since it can help create plans for animal management and conservation that work as well as stimulate new ideas in the fields of artificial intelligence and optimization algorithms. Our goal in this work is to explore the fascinating realm of prey–predator systems and learn more about the nuances of cooperative hunting, herd behavior, and the Holling type-II functional response. We seek to improve conservation science, deepen our understanding of ecological dynamics, and stimulate more research in this fascinating topic by illuminating these dynamic relationships.

The study of the dynamics of a single population was the original subject of mathematical population theory, which has its origins in the work of economist Thomas Malthus [1] from two centuries ago. In order to solve the flaws of the original model, such as the difficulty of unrestricted population increase given the available resources, the so-called logistic variation was later devised [2]. Thanks to the influence of Lotka and Volterra's studies [4, 5], scientists began to consider interacting populations after more than a century, especially of the competitive variety [3]. Cooperative hunting, or the increase in successful prey capture observed when multiple individuals participate in a hunt, has been demonstrated by a variety of animals. Many cooperatively hunting animals' hunts are best described as emergent, simultaneous individual hunts, in which each animal aims to maximize its chances of capturing the prey for itself. According to the role differentiation theory, people will choose to play roles that entail a higher risk of harm or a lower chance of success than they would in other roles. For example, during hunts, some people will react as chasers while others will block the prey's paths [6]. To describe cooperative hunting in the predator and group defense in the prey, the appropriate functional response to predation should be put forth [7].

Many complex occurrences have been seen, and Holling type-I and type-II functional responses have been widely addressed to explain predator–prey predation [8, 9, 10, 11, 12]. In order to understand group defense in the prey, recent research has concentrated on creating functional responses [13, 14, 15, 16, 17]. Using a square root functional response and a traditional Lotka–Volterra interaction term, Braza [18] investigated the dynamics of a predator–prey system. Freedman and Wolkowicz [19] investigated the effect of group defense by creating a nonmonotone predator response function that is monotonically increasing up to a threshold density and then monotonically decreasing. Certain prey species organize into herds, with the weaker members residing in the center of the herd and the stronger members on the outside. This indicates that predator attacks mostly target the animals close to the herd's edge. There are intricate and nonlinear ecological models of how populations interact in the natural world [20]. Numerous scholars have examined the intricate dynamics of the food web and predator–prey food chains; see, for instance, [21–24]. May [25] introduced chaos and chaotic dynamics in a basic model identified by the logistic map and demonstrated chaotic dynamics in an unforeseen way. It is challenging to create a mathematical model for ecology and biology since a mathematical model could never fully predict the behavior of nature. An analyzable model that is close enough to reality could be sought after to deal with this important problem. For instance, in many species' ecology models of prey–predator relationships, bifurcation diagrams and numerical simulation have been used to identify unanticipated structure. The functional response of a prey–predator model was developed by Lotka and Volterra in 1925 and 1926. However, as demonstrated in [25] and [26], appropriate functional responses for examining the complexity of model ecosystems must be limited and nonlinear. In 1959, Holling [27] used Michaelis–Menten's formula $\frac{wx}{c+\mu}$ to represent the functional responses of predators. This is referred to as a Michaelis–Menten functional reaction or Holling type-II response.

Recently, a few studies [28, 29, 30] have considered a particular kind of predator–prey system that exhibits herd behavior. Prey species gather together in herds to protect themselves from predators in these kinds of systems, such as those involving herbivores on large savannas and their large carnivores. Predators interact with prey along the outside corridor of the herd of prey. This will make the formulation of the prey response functions in terms of the square root of the number of prey species for that class of systems quite sensible. Numerous articles [31, 32, 33] address how the shape of mortality terms affects dynamics. Recently, Yuan et al. [31] studied a spatial predator–prey model with herd behavior and concluded that the Turing pattern is caused by quadratic mortality. While the "linear form" is more frequently employed to characterize the mortality term of predators (like piscivore), Brentnall et al. [32] state that the quadratic mortality is adequate for intermediate predators. Fulton et al. [33] state that in the wide, linked webs used in ecosystem models, model behavior is more sensitive to the form of the grazing term than mortality factors that are close to the modeled food web. Sanling Yuan [34] and Chaoqun Xu investigated the prey–predator model in which the prey population exhibits herd behavior. In their study, Singh and Banerjee [35] examined a spatial predator–prey model in which the predators engaged in cooperative hunting. Using linear stability analysis, the appropriate domain in the space of governing parameters was found, and the condition for diffusive instability was obtained. Authors can create complex patterns in the Turing domain, including the stripe pattern, spotted pattern, and stripe–spot (mixed) pattern, by varying the hunting cooperation parameter in predators and the carrying capacity of prey. The results, which focus on the influence of hunting cooperation in pattern dynamics of a diffusive predator–prey model, provide us with a good understanding of the structures of the predator–prey interactivity in real ecosystems. If prey display herd behavior, hunting cooperation is an important component of the biological system [36]. Based on the previously reviewed research, we build a new prey–predator model in this work by fusing cooperative behavior in the predators and herd behavior in the prey with a Holling type-II functional response in the prey–predator system. Our primary goal in this essay is to examine and study the temporal and spatial dynamics of the prey–predator system. The effects of cooperative hunting and the geographic patterns brought about by variations in the diffusion coefficient, K, and cooperation hunting rate (α) are of special interest to us.

The chapter is organized as follows. In the next section, we show how the model was derived. Before beginning the model analysis, we first explore the temporal model and demonstrate the positivity and boundedness of the solutions. Section 10.3 deals with the occurrence of equilibrium points and an analysis of their local stability. By changing the values of the parameters α and K, the Hopf bifurcation analysis of the system may be found in Section 10.4. We examine the effects of species dispersal on time and space in Section 10.5 by integrating diffusion into the temporal model. Section 10.6 yields diffusion-driven instability. The amplitude equation was derived in Section 10.7. On the other hand, the numerical simulation that demonstrates our analytical conclusions is presented in Section 10.8. We provide a concise analysis and summary of the main conclusions in the last section.

10.2 Mathematical Model Formulation and Preliminaries

Combining cooperative conduct among predators with herd behavior in prey, along with the Holling type-II functional response in the prey–predator system, results in the following predator–prey model:

$$\begin{cases} \dfrac{dX}{dT'} = gX(1 - \dfrac{X}{C}) - \dfrac{(l' + aY)\sqrt{X}Y}{1 + h_1(l' + aY)\sqrt{X}}, \\ \dfrac{dY}{dT'} = \dfrac{e(l' + aY)\sqrt{X}Y}{1 + h_1(l' + aY)\sqrt{X}} - m'Y. \end{cases} \tag{10.1}$$

At time T', let $X(T')$ represent the density of prey and $Y(T')$ represent the density of predators. In this case, the prey's intrinsic growth rate is g, and its carrying capacity is C. We investigate a Holling type-II functional response that looks like this:

$$\frac{(l' + aY)\sqrt{X}}{1 + h_1(l' + aY)\sqrt{X}}. \tag{10.2}$$

This demonstrates hunting cooperation and is dependent on both predator and prey populations. Prey's herd behavior is represented by the word $\sqrt{X}$. The parameters $a(a > 0)$ define predator cooperation in hunting (aY is the cooperation term), the parameter l' reflects the attack rate per predator and prey, and the parameter h_1 represents the predator's handling time with a prey. e represents the conversion efficiency, and m' is the per capita predator mortality rate. The addition of the dimensionless variables has made the model (10.1) non-dimensional.

$$\mu = \frac{el'X}{m'}, \ \vartheta = \frac{l'Y}{m'}, \ t = m'T', \tag{10.3}$$

and dimensionless parameters

$$\sigma = \frac{g}{m'}, K = \frac{el'C}{m'}, \alpha = \frac{am'}{l'^2}, h = \frac{m'}{e}h_1 \tag{10.4}$$

then get the updated model as

$$\begin{cases} \dfrac{d\mu}{dt} = \sigma\mu(1 - \dfrac{\mu}{K}) - \dfrac{\sqrt{\mu}\vartheta(1 + \alpha\vartheta)}{h(1 + \alpha\vartheta)\sqrt{\mu} + 1} = f(u, \vartheta), \\ \dfrac{d\vartheta}{dt} = \dfrac{\sqrt{\mu}\vartheta(1 + \alpha\vartheta)}{h(1 + \alpha\vartheta)\sqrt{\mu} + 1} - \vartheta = g(\mu, v) \end{cases} \tag{10.5}$$

with

$$\mu(0) > 0, \ \vartheta(0) > 0. \tag{10.6}$$

10.2.1 Existence and uniqueness

For the first-order ordinary nonlinear differential equation system with initial conditions, the existence and uniqueness are established. We have $U = [\mu, \vartheta]^T, F : R_2 \to R^2, F = [f, g]^T$, in the case of $t > 0$ and system (10.5), which may be expressed as $\frac{dU}{dt} = F(U)$. Here $f(\mu, \vartheta) = \sigma\mu(1 - \frac{\mu}{K}) - \dfrac{(1 + \alpha\vartheta)\sqrt{\mu}\vartheta}{1 + h(1 + \alpha\vartheta)\sqrt{\mu}} \in C^\infty(R^2)$ and $g(\mu, \vartheta) = \frac{(1+\alpha\vartheta)\sqrt{\mu}\vartheta}{1+h(1+\alpha\vartheta)\sqrt{\mu}} - \vartheta \in K^\infty(C^2)$. Now we can see that f is a smooth function for the variable (μ, ϑ) in R^2_+. Hence, the uniqueness and local existence of the solution are verified.

10.2.2 Positivity and boundedness

Theorem 10.1. Each solution of eqn (10.5) with respect to the initial conditions (10.6), which exists in $[0, \infty)$, remains positive $\forall t > 0$.

Proof. When we solve the following ordinary differential equation

$$\frac{d\mu}{dt} = \sigma\mu(1 - \frac{\mu}{K}) - \frac{\sqrt{\mu}\vartheta(1 + \alpha\vartheta)}{h(1 + \alpha\vartheta)\sqrt{u} + 1} \tag{10.7}$$

we get

$$\mu(t) = \mu(0)\exp\left(\int_0^t \left[\sigma\left(1 - \frac{\mu(\mathrm{r})}{\mathrm{K}}\right) - \frac{(1 + \alpha\vartheta(\mathrm{r}))\vartheta(\mathrm{r})}{\left(1 + \mathrm{h}(1 + \alpha\vartheta(\mathrm{r}))\sqrt{\mu(\mathrm{r})}\right)\sqrt{\mu(\mathrm{r})}}\right]\mathrm{dr}\right). \tag{10.8}$$

As we know, $\mu(0) > 0$; therefore, $\mu(t)$ is always positive for infinite time. In a similar way, by taking the second equation from system (10.5), we have

$$\vartheta(t) = \vartheta(0)\exp\left(\int_0^t \left[\frac{(1 + \alpha\vartheta(\mathrm{r}))\sqrt{\mu(\mathrm{r})}}{1 + \mathrm{h}(1 + \mathrm{av}(\mathrm{r}))\sqrt{\mu(\mathrm{r})}} - 1\right]\mathrm{dr}\right). \tag{10.9}$$

Thus, by using the condition $\vartheta(0) > 0$, we find that $\vartheta(t) > 0$ for all $t > 0$. Hence, R^2_+ must be positively invariant for the model (10.5). □

Theorem 10.2. Each solution of system (10.5) with initial conditions (10.6) is uniformly bounded in R^2_+.

Proof. With initial conditions 10.6, let $P = \mu(t) + \vartheta(t)$ and $(\mu(t), \vartheta(t))$ be any solution of system 10.5.

$$\dot{P} = \frac{d\mu}{dt} + \frac{d\vartheta}{dt} = \sigma\mu(1 - \frac{\mu}{K}) - \vartheta$$

$$\frac{dP}{dt} + \bar{\omega}P(t) = \mu(\sigma(1 - \frac{\mu}{K}) + \bar{\omega}) + (\bar{\omega} - 1)\vartheta.$$

Here, we select $\bar{\omega} \in (0, 1)$ and the maximum value of $\mu(\sigma(1 - \frac{\mu}{K}) + \bar{\omega})$ is $\frac{K(\bar{\omega}+\sigma)^2}{4\sigma}$.

Then we have

$$\frac{dP}{dt} + \bar{\omega}P(t) \leq \frac{K(\bar{\omega} + \sigma)^2}{4\sigma} = S. \tag{10.10}$$

On applying the Gronwall lemma [37], we get

$$0 < P(t) \leq \frac{S(1 - \exp(-\bar{\omega}\mathrm{t}))}{\bar{\omega}} + P(0)\exp(-\bar{\omega}\mathrm{t})).$$

Now, if $t \to \infty$, we obtain $0 < P(t) \leq \frac{S}{\bar{\omega}}$. Hence proved. □

10.3 Linear Stability Analysis for Non-spatial Model

There are three equilibrium points given by $E_I(0,0), E_{II}(K,0)$, and $E' = (\mu^*, \vartheta^*)$. Here $\mu^* = \frac{1}{(1+\alpha\vartheta)^2(1-h)^2}$ and ϑ^* is the solution of the equation

$$\beta_1\vartheta^5 + \beta_2\vartheta^4 + \beta_3\vartheta^3 + \beta_4\vartheta^2 + \beta_5\vartheta + \beta_6 = 0,$$

where

$$\begin{aligned}
\beta_1 &= -\alpha^4 K(1-h)^4,\\
\beta_2 &= -4\alpha^3 K(1-h)^4,\\
\beta_3 &= \alpha^3\sigma K(1-h)^2 - 6\alpha^2 K(1-h)^4,\\
\beta_4 &= 3\alpha^2\sigma K(1-h)^2 - 4\alpha K(1-h)^4,\\
\beta_5 &= \alpha\sigma(3K(1-h)^2 - 1) - K(1-h)^4,\\
\beta_6 &= \sigma(K(1-h)^2 - 1).
\end{aligned}$$

The nullclines for various values of hunting cooperation (α) are presented in Figure 10.1 and it is obvious that the system 10.5 has no coexistence equilibrium if the cooperation among predators is lower.

The variational matrix for model 10.5 is given by

$$V_R = \begin{bmatrix} \dfrac{-(\alpha\vartheta+1)\vartheta}{2\sqrt{\mu}K_1^2} - \dfrac{2\sigma\mu}{K} + \sigma & \dfrac{-h\mu(\alpha\vartheta+1)^2 - \sqrt{\mu}(2\vartheta\alpha+1)}{(K_1)^2} \\ \dfrac{(\alpha\vartheta+1)\vartheta}{2\sqrt{\mu}K_1^2} & \dfrac{h\mu(\alpha\vartheta+1)^2 + \sqrt{\mu}(2\vartheta\alpha+1)}{(K_1)^2} - 1 \end{bmatrix},$$

where $K_1(\mu,\vartheta) = (h\sqrt{\mu}(\alpha\vartheta+1)+1)$.

Theorem 10.3. At the equilibrium point $E_{II} = (K,0)$, system 10.5 is locally asymptotically stable if $K > \left(\frac{hK(1-h)-1}{(2h-1)}\right)^2$ and not stable if $K < \left(\frac{hK(1-h)-1}{(2h-1)}\right)^2$.

Proof. The variational matrix V_R at $E_{II} = (K,0)$ is $V_{II} = \begin{bmatrix} -\sigma & \frac{-hK-\sqrt{K}}{(h\sqrt{K}+1)^2} \\ 0 & \frac{hK+\sqrt{K}}{(h\sqrt{K}+1)^2} - 1 \end{bmatrix}$.

Its eigenvalues are $-\sigma$ and $\frac{\sqrt{K}}{K_1} - 1$. If eigenvalues of V_{II} has a negative real part, then the axial equilibrium will be stable. Here each parameter value is positive; therefore, the system is stable if $K > \left(\frac{hK(1-h)-1}{(2h-1)}\right)^2$. □

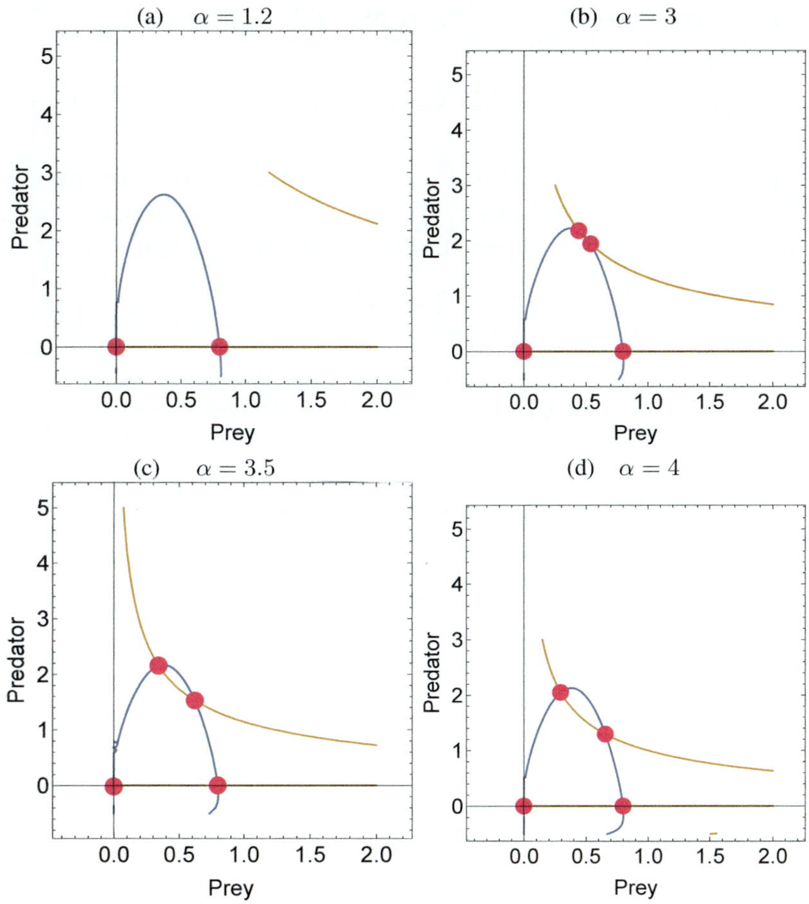

Figure 10.1 Illustration of equilibrium points and nullclines of system 10.5 with an increase in the value of α. Equilibrium points are represented by light blue dots. Fixed values of the parameters are $\sigma = 11$ and $h = 0.8$, $K = 0.8$.

A portion of the environment and the species living there are described as an ecosystem, together with all the interactions and relationships that imply. Due to these interactions, it becomes more important to check the stability of the system at an equilibrium where both prey and predator populations exist, i.e., $E' = (\mu^*, \vartheta^*)$. Now the variational matrix at $E' = (\mu^*, \vartheta^*)$ is

$$V_R^* = \begin{bmatrix} f_\mu & f_\vartheta \\ g_\mu & g_\vartheta \end{bmatrix}_{(\mu^*, \vartheta^*)}$$

$$= \begin{bmatrix} \frac{-(\alpha\vartheta^*+1)\vartheta^*}{2\sqrt{\mu^*}K_2^2} - \frac{2\sigma\mu^*}{K} + \sigma & \frac{-h\mu^*(\alpha\vartheta^*+1)^2 - \sqrt{\mu^*}(2\vartheta^*\alpha+1)}{(K_2)^2} \\ \frac{(\alpha\vartheta^*+1)\vartheta^*}{2\sqrt{\mu^*}K_2^2} & \frac{h\mu^*(\alpha\vartheta^*+1)^2 + \sqrt{\mu^*}(2\vartheta^*\alpha+1)}{(K_2)^2} - 1 \end{bmatrix}.$$

where $K_2(\mu^*, \vartheta^*) = (h\sqrt{\mu^*}(\alpha\vartheta^* + 1) + 1)$. The Routh–Hurwitz criterion states that if Tr $(V_R^*) < 0$ and Det $(V_R^*) > 0$, then the eigenvalues of eqn (10.11) will have a negative real part. The characteristic equation of V_R^* is

$$\lambda^2 - \mathrm{Tr}(\mathrm{V}_\mathrm{R}^*)\lambda + \mathrm{Det}(\mathrm{V}_\mathrm{R}^*) = 0, \tag{10.11}$$

where

$$\begin{aligned} \mathrm{Tr}(\mathrm{V}_\mathrm{R}^*) &= f_\mu + g_\vartheta, \\ \mathrm{Det}(\mathrm{V}_\mathrm{R}^*) &= f_\mu g_\vartheta - g_\mu f_v. \end{aligned} \tag{10.12}$$

The conditions obtained from the above study are written in the form of a theorem as follows.

Theorem 10.4. At $E' = (\mu^*, \vartheta^*)$, model (10.11) becomes asymptotically stable when $K > \left(\frac{hK(1-h)-1}{(2h-1)}\right)^2$, and also the below given inequalities are satisfied

(1) $\sigma + \frac{h\mu^*(\alpha\vartheta^*+1)^2+\sqrt{\mu^*}(2\vartheta^*\alpha+1)}{(K_2)^2} < 1 + \frac{(\alpha\vartheta^*+1)\vartheta^*}{2\sqrt{\mu^*}K_2^2} + \frac{2\sigma\mu^*}{K}$,

(2) $\frac{\left(-(\alpha\vartheta^*+1)\vartheta^*K - 4\sigma\mu^{*3/2}K_2^2 + 2\sqrt{\mu^*}K_2^2K\sigma\right)\left(h\mu^*(\alpha\vartheta^*+1)^2+\sqrt{\mu^*}(2\vartheta^*\alpha+1)-K_2^2\right)}{2\sqrt{\mu^*}K_2^4K}$

$> \frac{\left((\alpha\vartheta^*+1)\vartheta^*\right)\left(-h\mu^*(\alpha\vartheta^*+1)^2-\sqrt{\mu^*}(2\vartheta^*\alpha+1)\right)}{2\sqrt{\mu^*}K_2^4}$.

10.4 Hopf Bifurcation

In this section, we perform the bifurcation investigation exclusively for the coexistence equilibrium point $E' = (\mu^*, \vartheta^*)$, using K as a bifurcation parameter. It is commonly known that the existence of a non-hyperbolic equilibrium point is a required but not sufficient condition for bifurcation. Hopf bifurcation occurs at E' when the linearized model has guaranteed complex conjugate eigenvalue pairs with imaginary portion alone.

After solving eqn (10.11), we have observed that it has eigenvalues, which are given by

$$\jmath_\imath = \frac{1}{2}(P_1 \pm \sqrt{P_1^2 - 4P_2}), \imath = 1, 2, \tag{10.13}$$

where both $P_1 = \mathrm{Tr}(\mathrm{V}^*_{\mathrm{R}})$ and $P_2 = \mathrm{Det}(\mathrm{V}^*_{\mathrm{R}})$ are the functions of K. If $P_1 = 0$ and $P_2 > 0$, the eigenvalues (10.11) become pure complex eigenvalues. Now if we considered bifurcation parameter as K, then the Hopf bifurcation threshold value $\bar{K}$ can be calculated by setting $K = \bar{K}$ while keeping other parameters the same in $P_1 = 0$, and we get

$$\bar{K} = \frac{4\sigma\mu^{*\frac{3}{2}}K_2^2}{2\sqrt{\mu^*}K_2^2(\sigma-1)+(\alpha\vartheta^*+1)(2u^*K_2-\vartheta^*)+2\alpha\mu^*v^*}. \quad (10.14)$$

Theorem 10.5. At the equilibrium point $E' = (\mu^*, \vartheta^*)$, Hopf bifurcation exist for system (10.5), if

(1) $Q_1(\bar{K}) = 0$, and $Q_2(\bar{K}) \neq 0$.

(2) $\left[\frac{d}{dK}(\mathrm{Tr}(\mathrm{V}^*_{\mathrm{R}})\right]_{\mathrm{K}=\bar{\mathrm{K}}} \neq 0$.

Proof. Let $\lambda(\bar{K}) = Q_1(K) + \iota Q_2(K)$ be the complex eigenvalues of the characteristic polynomial (10.11). Now put the value of λ in eqn (10.11), and take the real and imaginary parts equal to zero separately; so we get

$$Q_1^2 - Q_2^2 + P_1Q_1 + P_2 = 0 \text{ and } 2Q_1Q_2 + P_1Q_2 = 0. \quad (10.15)$$

Choosing $K = \bar{K}$ such that $Q_1(\bar{K}) = 0$, and substituting to eqn (10.15), we get

$$-Q_2^2 + P_2 = 0, \text{ and } P_1Q_2 = 0 \text{ with } Q_2 \neq 0. \quad (10.16)$$

From the above equation, we have $P_1(\bar{K}) = 0$ and $Q_2(\bar{K}) = \sqrt{P_2(\bar{K})}$. This gives $\lambda(\bar{K}) = -\iota\sqrt{P_2(\bar{K})}$. Now differentiating the two equations in (10.15) with respect to K, and substituting $Q_1(\bar{K}) = 0$, we have

$$\begin{aligned} P_1\frac{dQ_1(K)}{dK} - 2Q_2\frac{dQ_2(K)}{dK} &= -\frac{dP_2(K)}{dK}, \\ 2Q_2\frac{dQ_1(K)}{dK} + P_1\frac{dQ_2(K)}{dK} &= -Q_2\frac{dP_1(K)}{dK}. \end{aligned} \quad (10.17)$$

On solving eqn (10.17), we get

$$\left[\frac{d}{dK}(\mathrm{Tr}(\mathrm{V}^*_{\mathrm{R}}))\right]_{\mathrm{K}=\bar{\mathrm{K}}} = -\left[\frac{\mathrm{P}_1\frac{\mathrm{P}_2}{\mathrm{dK}} + 2\mathrm{Q}_2^2\frac{\mathrm{P}_1}{\mathrm{dK}}}{\mathrm{P}_1 + 4\mathrm{Q}_2^2}\right]_{\mathrm{K}=\bar{\mathrm{K}}} \neq 0, \quad (10.18)$$

provided $P_1\frac{P_2}{dK} + 2Q_2^2\frac{P_1}{dK} \neq 0$. □

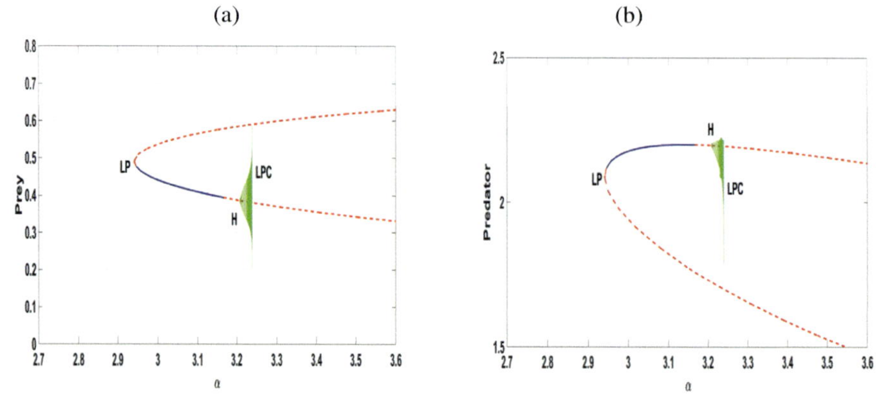

Figure 10.2 Diagram showing Hopf bifurcation for the parameter α.

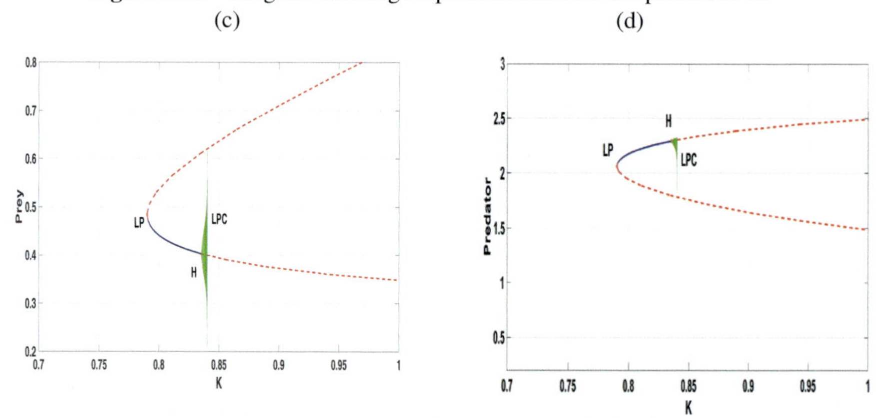

Figure 10.3 Diagram showing Hopf bifurcation for the parameter K.

10.5 Spatial Model

Taking into account the species' random movements as they seek out of shelter, food and safety from the predators along with the addition of diffusion in system 10.5. Relative to eqn 10.5, the spatial model is given by

$$\begin{cases} \frac{\partial\mu}{\partial t} = \sigma\mu(1-\frac{\mu}{K}) - \dfrac{(1+\alpha\vartheta)\sqrt{\mu}\vartheta}{1+h(1+\alpha\vartheta)\sqrt{\mu}} + d_\mu\nabla^2\mu, \\ \frac{\partial\vartheta}{\partial t} = \dfrac{(1+\alpha\vartheta)\sqrt{\mu}\vartheta}{1+h(1+\alpha\vartheta)\sqrt{\mu}} - \vartheta + d_\vartheta\nabla^2\vartheta, \end{cases} \tag{10.19}$$

where $\nabla^2 = \frac{\partial^2}{\partial\mu^2} + \frac{\partial^2}{\partial\vartheta^2}$, and d_ϑ is the diffusion coefficients for the prey and d_μ for predator population. Both d_μ and d_ϑ are non-negative. We take the zero flux boundary and initial condition as given below:

$$\mu(\mu_1, \mu_2, 0) > 0,\ \vartheta(\mu_1,\mu_2, 0) > 0,\ (\mu_1, \mu_2) \in \Omega = [0, M] \times [0, M],$$
$$\frac{\partial\mu}{\partial\eta} = 0 = \frac{\partial\vartheta}{\partial\eta}. \tag{10.20}$$

where M represents the size of the system in the two directions U and V, and the unit normal vector at the boundary domain $\partial\Omega$ in the outward direction is denoted by η.

10.6 Diffusion-driven Instability

In the current portion, we examined the stability of spatial model 10.19 and obtained the diffusion-driven instability condition. First of all, by using Taylor series expansion, system (10.19) is linearized at the coexistence equilibrium point, as given below:

$$\dot{Z} = V_R^* Z + D\Delta Z, \tag{10.21}$$

where V_R^* stands for variational matrix given in eqn (10.11), and

$$Z = \begin{bmatrix} \mu - \mu_1 \\ \vartheta - \vartheta_1 \end{bmatrix} \text{ and } D = \begin{bmatrix} d_\mu & 0 \\ 0 & d_\vartheta \end{bmatrix}.$$

Now, let us suppose that the solution to (10.21) can be written as

$$Z = \epsilon c_k \exp\left(\lambda t + i\mathbf{k.r}\right) + \text{c.c.} + \mathrm{O}(\epsilon^2), \tag{10.22}$$

where λ is the growth rate of perturbation in time t, i(iota) denotes the imaginary unit number and the value of i is $\sqrt{-1}$, c.c. stands for the complex conjugate, **k** is the wave number, and **r** is the spatial vector.

On substituting eqn (10.22) into eqn (10.19), we get the characteristic equation as

$$\begin{vmatrix} f_\mu - k^2 d_\mu - \lambda & f_\vartheta \\ g_\mu & g_\vartheta - k^2 d_\vartheta - \lambda \end{vmatrix} = 0. \tag{10.23}$$

From eqn (10.23), we obtained a characteristic equation:

$$\lambda^2 - \varrho_1(k^2)\lambda + \varrho_2(k^2) = 0, \tag{10.24}$$

where

$$\begin{aligned} \varrho_1(k^2) &= \mathrm{Tr}(\mathrm{V}_{\mathrm{R}}^*) - \mathrm{k}^2(\mathrm{d}_\mu + \mathrm{d}^2), \\ \varrho_2(k^2) &= d_\mu d_\vartheta k^4 - (d_\vartheta f_\mu + d_\mu f_\vartheta)k^2 + \mathrm{Det}(\mathrm{V}_{\mathrm{R}}^*). \end{aligned} \tag{10.25}$$

Now, the system is stable at E' by using Theorem 10.4. While the diffusion factor will be induced, diffusion-driven instability in the diffusive model, that is E', will become unstable. Here, $\mathrm{Tr}(V_R^*) < 0$, $d_\mu > 0$ and $d_\vartheta > 0$; thus, $\varrho_1(k^2) < 0$. Therefore, the system is not stable, i.e., $\varrho_2(k^2) < 0$.

Hence, the diffusion-driven instability will be given by

$$d_\mu d_\vartheta k^4 - (d_\vartheta f_\mu + d_\mu f_\vartheta)k^2 + \mathrm{Det}(\mathrm{V}_{\mathrm{R}}^*) < 0. \tag{10.26}$$

For some $k = k_T$, $\varrho_2(k^2) = 0$, and $k_{min}^2 = \frac{d_\vartheta f_\mu + d_\mu f_\vartheta}{2 d_\mu d_\vartheta}$ is the minimum value of k. Then

$$\Gamma_2(k_{\mathrm{min}}^2) = \mathrm{Det}(\mathrm{V}_{\mathrm{R}}^*) - \frac{(\mathrm{d}_\vartheta \mathrm{f}_\mu + \mathrm{d}_\mu \mathrm{f}_\vartheta)^2}{4\mathrm{d}_\mu \mathrm{d}_\vartheta}, \tag{10.27}$$

and the wave number has the critical value

$$k^2 = k_T^2 = \sqrt{\frac{\mathrm{Det}(\mathrm{V}_{\mathrm{R}}^*)}{d_\mu d_\vartheta}}. \tag{10.28}$$

On putting eqn (10.28) in $\Gamma_2(k^2)$, we identify the required criteria for diffusion-driven instability to arise as

$$(d_\vartheta f_\mu + d_\mu f_\vartheta)^2 > 4(f_\mu g_\vartheta - g_\mu f_\vartheta) d_\mu d_{(2)}. \tag{10.29}$$

Now, we have a theorem on the basis of the above study.

Theorem 10.6. At E', system (10.19) displays diffusion-driven instability if

(1) $f_\mu + g_\vartheta < 0$,
(2) $f_\mu g_\vartheta - g_\mu f_v > 0$,
(3) $d_\vartheta f_\mu + d_\mu f_\vartheta > 0$,
(4) $(d_\vartheta f_\mu + d_\mu f_\vartheta)^2 > 4(f_\mu g_\vartheta - g_\mu f_\vartheta) d_\mu d_\vartheta$.

Next, we formulate the amplitude equations using d_ϑ as the bifurcation parameter. The Turing region that corresponds to the bifurcation parameter $d_\vartheta = {d_\vartheta}^T$, for the Turing pattern orientation in Figure 10.4.

10.7 Weakly Nonlinear Analysis

In this section, the 2D spatial patterns to the 2D reaction–diffusion system are examined using the weakly nonlinear stability technique that was first presented by Segel and Levin by Waltman [3] and Wollkind and Stephenson by An and Jiang [13]. Near the emergence $\rho = \rho^T$, the equilibria become unstable, the important modes' eigenvalues approach zero, and these modes undergo progressive changes, while the off-critical modes slack suddenly. It follows that only interruptions with k near k^T require attention. Consequently, the spatial dynamics could be replaced by the dynamics of dynamic dormant modes. We are going to obtain the amplitude equations by employing the multiple-scale approach. First, we must use Taylor's series to linearize system (10.19) in order to derive the amplitude equations.

$$\frac{\partial U}{\partial t} = MU + J, \tag{10.30}$$

where

$$\begin{aligned}
U &= \begin{bmatrix} \mu \\ \vartheta \end{bmatrix}, \; M = \begin{bmatrix} f_\mu + d_\mu \nabla^2 & f_\vartheta \\ g_\mu & g_\vartheta + d_\vartheta \nabla^2 \end{bmatrix}, \text{ and} \\
J &= \epsilon^2 J_2 + \epsilon^3 J_3 + O(\epsilon^4), \\
J_2 &= \frac{1}{2}\begin{bmatrix} f_{\mu} x_1^2 + 2 f_{\mu\vartheta}\mu_1\vartheta_2 + f_{\vartheta\vartheta}\vartheta_2^2 \\ g_{\mu\mu} x_1^2 + 2 g_{\mu\vartheta}\mu_1\vartheta_2 + g_{\vartheta\vartheta}\vartheta_2^2 \end{bmatrix}, \\
J_3 &= \begin{bmatrix} f_\mu \mu_1\mu_2 + f_{\mu\vartheta}(\mu_1\vartheta_2 + \mu_2\vartheta_2) + f_{\vartheta\vartheta}\vartheta_2\vartheta_2 + \frac{1}{6}\Big(f_{\mu\mu\mu}x_1^3 \\ +3 f_{\mu\mu\vartheta}x_1^2\vartheta_2 + 3 f_{\mu\vartheta\vartheta}\mu_1\vartheta_2^2 + f_{\vartheta\vartheta\vartheta}\vartheta_2^3\Big) \\ g_{\mu\mu}\mu_1\mu_2 + g_{\mu\vartheta}(\mu_1\vartheta_2 + \mu_2\vartheta_2) + g_{\vartheta\vartheta}\vartheta_2\vartheta_2 + \frac{1}{6}\Big(g_{\mu\mu\mu}x_1^3 \\ +3 g_{\mu\mu\vartheta}x_1^2\vartheta_2 + 3 g_{\mu\vartheta\vartheta}\mu_1\vartheta_2^2 + g_{\vartheta\vartheta\vartheta}\vartheta_2^3\Big) \end{bmatrix},
\end{aligned} \tag{10.31}$$

where the nonlinear and linear terms are denoted by J and M, respectively. Now, expanding the bifurcation parameter d_ϑ and U with the application of perturbation theory in system (10.19), we get

$$\begin{aligned}
d_\vartheta^T - d_\vartheta &= \epsilon d_\vartheta^{(1)} + \epsilon^2 d_\vartheta^{(2)} + O(\epsilon^3), \\
U = \begin{bmatrix} \mu \\ \vartheta \end{bmatrix} &= \epsilon \begin{bmatrix} \mu_1 \\ \vartheta_2 \end{bmatrix} + \epsilon^2 \begin{bmatrix} \mu_2 \\ \vartheta_2 \end{bmatrix} + \epsilon^3 \begin{bmatrix} \mu_3 \\ \vartheta_3 \end{bmatrix} + O(\epsilon^4).
\end{aligned} \tag{10.32}$$

On putting the value of $(d_\vartheta^T - d_\vartheta)$ from eqn 10.32 in M, we obtain

$$M = M_T + \epsilon d_\vartheta^{(1)} H + \epsilon^2 d_\vartheta^{(2)} H + O(\epsilon^3), \tag{10.33}$$

where

$$M_T = \begin{bmatrix} f_\mu + d_\mu \nabla^2 & f_\vartheta \\ g_\mu & g_\vartheta + d_\vartheta^T \nabla^2 \end{bmatrix} \text{ and } H = \begin{bmatrix} 0 & 0 \\ 0 & \nabla^2 \end{bmatrix}. \tag{10.34}$$

The system's time scale must now be distributed so that each time scale t_i can be viewed as an independent variable. The time derivative should translate into the following phrases according to the derivative chain rule:

$$\frac{\partial}{\partial t} = \epsilon \frac{\partial}{\partial t_1} + \epsilon^2 \frac{\partial}{\partial t_2} + \epsilon^3 \frac{\partial}{\partial t_3} + O(\epsilon^4), \tag{10.35}$$

and expanding eqn (10.35) as per a distinct order of *E* results in $O(\epsilon)$ of first order:

$$M^T \begin{bmatrix} \mu_1 \\ \vartheta_2 \end{bmatrix} = 0, \tag{10.36}$$

second order $O(\epsilon^2)$:

$$M^T \begin{bmatrix} \mu_2 \\ \vartheta_2 \end{bmatrix} = \frac{\partial}{\partial t_1} \begin{bmatrix} \mu_1 \\ \vartheta_2 \end{bmatrix} - d_\vartheta^{(1)} H \begin{bmatrix} \mu_1 \\ \vartheta_2 \end{bmatrix} - J_2, \text{ and} \tag{10.37}$$

third order $O(\epsilon^3)$:

$$M^T \begin{bmatrix} \mu_3 \\ \vartheta_3 \end{bmatrix} = \frac{\partial}{\partial t_1} \begin{bmatrix} \mu_2 \\ \vartheta_2 \end{bmatrix} + \frac{\partial}{\partial t_2} \begin{bmatrix} \mu_1 \\ \vartheta_2 \end{bmatrix} - d_\vartheta^{(1)} H \begin{bmatrix} \mu_2 \\ \vartheta_2 \end{bmatrix} - d_\vartheta^{(2)} H \begin{bmatrix} \mu_1 \\ \vartheta_2 \end{bmatrix} - J_3. \tag{10.38}$$

As we know, $[\mu_1, \vartheta_2]^T$ denotes the linear combination of the eigenvectors that are corresponding to the zero eigenvalue. Now solving eqn (10.36), we get

$$\begin{bmatrix} \mu_1 \\ \vartheta_2 \end{bmatrix} = \begin{bmatrix} \bar{\mu} \\ 1 \end{bmatrix} \sum_{j=1}^{3} [T_j \exp(i\mathbf{k}_j.\mathbf{r})] + c.c., \quad j = 1, 2, 3, \tag{10.39}$$

where the eigenvector corresponding to the zero eigenvalue is given by $\bar{\mu} = \frac{f_\mu d_\vartheta^T - g_\vartheta d_\mu}{2 g_\mu d_\mu}$, $|k| = k_T$, $T_j =$ the amplitude corresponding to the $|\exp(i\mathbf{k}_j.\mathbf{r})|$.

By using the "Fredholm solvability criterion," the vector function of eqn (10.37) must be orthogonal with the zero eigenvector of M^{T^*} in order to guarantee the existence of a non-trivial solution to this equation. The M^{T^*}'s null character vectors are

$$\begin{bmatrix} 1 \\ \bar{\vartheta} \end{bmatrix} (\exp(-i\mathbf{k}_j.\mathbf{r})) + c.c., \quad j = 1, 2, 3, \tag{10.40}$$

where $\bar{\vartheta} = \frac{2f_\vartheta d_\mu}{f_\mu d_\vartheta^T - g_\vartheta d_\mu}$. Now on the basis of condition of orthogonality, we get

$$\begin{bmatrix} 1 \\ \bar{\vartheta} \end{bmatrix} (\exp(-i\mathbf{k}_j.\mathbf{r}))J_2 = 0. \quad j = 1, 2, 3. \tag{10.41}$$

From eqn (10.41), we have

$$\begin{aligned}
(\bar{\mu} + \bar{\vartheta})\frac{\partial T_1}{\partial t_1} &= -k_T^2 d_\vartheta^{(1)} \bar{\vartheta} T_1 + ((f_{\mu\mu}\bar{\mu}^2 + f_{\mu\vartheta}\bar{\mu} + f_{\vartheta\vartheta}) \\
&\quad + \bar{\vartheta}(g_{\mu\mu}\bar{\mu}^2 + g_{\mu y}\bar{\mu} + g_{\vartheta\vartheta}))\bar{T}_2\bar{T}_3, \\
(\bar{\mu} + \bar{\vartheta})\frac{\partial T_2}{\partial t_1} &= -k_T^2 d_\vartheta^{(1)} \bar{\vartheta} T_2 + ((f_{\mu\mu}\bar{\mu}^2 + f_{\mu\vartheta}\bar{\mu} + f_{\vartheta\vartheta}) \\
&\quad + \bar{\vartheta}(g_{\mu\mu}\bar{\mu}^2 + g_{\mu\vartheta}\bar{\mu} + g_{\vartheta\vartheta}))\bar{T}_1\bar{T}_3, \\
(\bar{\mu} + \bar{\vartheta})\frac{\partial T_3}{\partial t_1} &= -k_T^2 d_\vartheta^{(1)} \bar{\vartheta} T_3 + ((f_{\mu\mu}\bar{\mu}^2 + f_{\mu\vartheta}\bar{\mu} + f_{\vartheta\vartheta}) \\
&\quad + \bar{\vartheta}(g_{\mu\mu}\bar{\mu}^2 + g_{\mu\vartheta}\bar{\mu} + g_{\vartheta\vartheta}))\bar{T}_1\bar{T}_2.
\end{aligned} \tag{10.42}$$

Now, substituting eqn (10.39) in eqn (10.37) and writing $[\mu_2 \ \ \vartheta_2]$ in the form of Fourier series, we get:

$$\begin{aligned}
\begin{bmatrix} \mu_2 \\ \vartheta_2 \end{bmatrix} &= \begin{bmatrix} \mu_{2(0)} \\ \vartheta_{2(0)} \end{bmatrix} + \sum_{j=1}^{3} \begin{bmatrix} \mu_{2(j)} \\ \vartheta_{2(j)} \end{bmatrix} \exp(i\mathbf{k}_j.\mathbf{r}) + \sum_{j=1}^{3} \begin{bmatrix} \mu_{2(ij)} \\ \vartheta_{2(ij)} \end{bmatrix} \exp(2i\mathbf{k}_j.\mathbf{r}) \\
&+ \begin{bmatrix} \mu_{2(12)} \\ \vartheta_{2(12)} \end{bmatrix} \exp(i(\mathbf{k}_1 - \mathbf{k}_2).\mathbf{r}) + \begin{bmatrix} \mu_{2(23)} \\ \vartheta_{2(23)} \end{bmatrix} \exp(i(\mathbf{k}_2 - \mathbf{k}_3).\mathbf{r}) \\
&+ \begin{bmatrix} \mu_{2(31)} \\ \vartheta_{2(31)} \end{bmatrix} \exp(i(\mathbf{k}_3 - \mathbf{k}_1).\mathbf{r}) + c.c,
\end{aligned} \tag{10.43}$$

where

$$\mu_{2(j)} = \bar{\mu}\vartheta_{2(j)}, \quad \begin{bmatrix} \mu_{2(0)} \\ \vartheta_{2(0)} \end{bmatrix} = \begin{bmatrix} x_{00} \\ y_{00} \end{bmatrix} (|T_1|^2 + |T_2|^2 + |T_3|^2),$$
$$\begin{bmatrix} \mu_{2(jj)} \\ \vartheta_{2(jj)} \end{bmatrix} = \begin{bmatrix} \mu_{11} \\ \vartheta_{11} \end{bmatrix} T_j^2, \quad \begin{bmatrix} \mu_{2(12)} \\ \vartheta_{2(12)} \end{bmatrix} = \begin{bmatrix} \mu_{22} \\ \vartheta_{22} \end{bmatrix} T_1\bar{T}_2, \tag{10.44}$$
$$\begin{bmatrix} \mu_{2(23)} \\ \vartheta_{2(23)} \end{bmatrix} = \begin{bmatrix} \mu_{22} \\ \vartheta_{22} \end{bmatrix} T_2\bar{T}_3, \quad \begin{bmatrix} \mu_{2(31)} \\ \vartheta_{2(31)} \end{bmatrix} = \begin{bmatrix} \mu_{22} \\ \vartheta_{22} \end{bmatrix} T_3\bar{T}_1,$$

with

$$\begin{bmatrix} x_{00} \\ y_{00} \end{bmatrix} = \begin{bmatrix} \frac{2f_\vartheta Q - 2g_\vartheta P}{f_\vartheta g_\mu - f_\mu g_\vartheta} \\ \frac{2(f_\mu Q - g_\mu P)}{f_\mu g_\vartheta - f_\vartheta g_\mu} \end{bmatrix}, \quad \begin{bmatrix} \mu_{11} \\ \vartheta_{11} \end{bmatrix} = \begin{bmatrix} \frac{-g_\vartheta P + f_\vartheta Q + 4d_\vartheta^T P k_T^2}{(4d_y^T k_T^2 - g_\vartheta)(f_\mu - 4d_\mu k_T^2) + f_\vartheta g_\mu} \\ \frac{g_\mu P - f_\mu Q + 4Q d_\mu k_T^2}{(4d_y^T k_T^2 - g_\vartheta)(f_\mu - 4d_\mu k_T^2) + f_\vartheta g_\mu} \end{bmatrix},$$

$$\begin{bmatrix} \mu_{22} \\ \vartheta_{22} \end{bmatrix} = \begin{bmatrix} \frac{-2g_\vartheta P + 2f_\vartheta Q + 6d_y^T P k_T^2}{(3d_y^T k_T^2 - g_\vartheta)(f_\mu - 3d_x k_T^2) + f_\vartheta g_\mu} \\ \frac{2g_\mu P - 2f_\mu Q + 6Q d_x k_T^2}{(3d_y^T k_T^2 - g_\vartheta)(f_\mu - 3d_x k_T^2) + f_\vartheta g_\mu} \end{bmatrix},$$

in which

$$P = -\left(\frac{1}{2} f_{\mu\mu}\bar{\mu}^2 + f_{\mu\vartheta}\bar{\mu} + \frac{1}{2} f_{\vartheta\vartheta}\right),$$
$$Q = -\left(\frac{1}{2} g_{\mu\mu}\bar{\mu}^2 + g_{\mu\vartheta}\bar{\mu} + \frac{1}{2} g_{\vartheta\vartheta}\right). \tag{10.45}$$

Using the solvability condition again in eqn (10.38), we have

$$(\bar{\mu} + \bar{\vartheta})\left(\frac{\partial Y_1}{\partial t_1} + \frac{\partial E_1}{\partial t_2}\right) = -k_T^2 d_\vartheta^T \bar{\vartheta}(d_\vartheta^{(2)} E_1 + d_\vartheta^{(1)} Y_1) + [f_{\mu\mu}\bar{\mu}^2 + 2f_{\mu\vartheta}\bar{\mu}$$
$$+ f_{\vartheta\vartheta} + \bar{\vartheta}(g_{\mu\mu}\bar{\mu}^2 + 2g_{\mu\vartheta}\bar{\mu} + g_{\vartheta\vartheta})](\bar{E}_2\bar{Y}_3 \tag{10.46}$$
$$+ \bar{E}_3\bar{Y}_2) - (G_1|E_1|^2 + G_2(|E_2|^2 + |E_3|^2)E_1).$$

The transformation of the subscript of E results in the other two equations. Now, μ and ϑ for the reaction–diffusion system (10.19) at the Turing instability's critical point have the form

$$\begin{bmatrix} \mu \\ \vartheta \end{bmatrix} = \sum_{j=1}^{3} (W_j \exp(i\mathbf{k}_j.\mathbf{r}) + \bar{W}_j \exp(i\mathbf{k}_j.\mathbf{r})), \tag{10.47}$$

where the associated amplitudes of modes k_j and $-k_j$ are W_j and $\bar{W}_j$, respectively; so we get

$$W_j = \epsilon T_j + \epsilon^2 Y_j + O(\epsilon^3) \quad j = 1, 2, 3. \tag{10.48}$$

This leads to

$$\begin{aligned}
\Upsilon_0 \frac{\partial W_1}{\partial t} &= \rho W_1 + \ell \bar{W}_2 \bar{W}_3 - [s_1 |W_1|^2 + s_2(|W_2|^2 + |W_3|^2)] W_1, \\
\Upsilon_0 \frac{\partial W_2}{\partial t} &= \rho W_2 + \ell \bar{W}_1 \bar{W}_3 - [s_1 |W_2|^2 + s_2(|W_3|^2 + |W_1|^2)] W_2, \\
\Upsilon_0 \frac{\partial W_3}{\partial t} &= \rho W_3 + \ell \bar{W}_1 \bar{W}_2 - [s_1 |W_3|^2 + s_2(|W_1|^2 + |W_2|^2)] W_3,
\end{aligned} \tag{10.49}$$

where

$$\begin{aligned}
\rho &= \frac{d_\vartheta^T - d_\vartheta}{d_\vartheta^T}, \quad \Upsilon_0 = \frac{(\bar{\mu} + \bar{\vartheta})}{-k_T^2 d_\vartheta^T \bar{\vartheta}}, \quad \ell = \frac{2(P + \bar{\vartheta} Q)}{-k_T^2 d_\vartheta^T \bar{\vartheta}}, \\
s_1 &= \frac{G_1}{-k_T^2 d_\vartheta^T \bar{\vartheta}}, \quad s_2 = \frac{G_2}{-k_T^2 d_\vartheta^T \bar{\vartheta}}.
\end{aligned} \tag{10.50}$$

$$\begin{aligned}
G_1 &= -\Big([(\bar{\mu}_1 f_{\mu\mu} + f_{\mu\vartheta})(\mu_{2(0)} + \mu_{2(11)}) + (f_{\vartheta\vartheta} + \bar{\mu}_1 f_{\mu\vartheta})(\vartheta_{2(0)} + \vartheta_{2(11)}) + R] \\
&\quad + \bar{\vartheta}[(\bar{\mu}_1 g_{\mu\mu} + g_{\mu\vartheta})(\mu_{2(0)} + \mu_{2(11)}) + (g_{\vartheta\vartheta} + \bar{\mu}_1 g_{\mu\vartheta})(\vartheta_{2(0)} + \vartheta_{2(11)}) + S] \Big), \\
G_2 &= -\Big([(\bar{\mu}_1 f_{\mu\mu} + f_{\mu\vartheta})(\mu_{2(0)} + \mu_{2(22)}) + (f_{\vartheta\vartheta} + \bar{\mu}_1 f_{\mu\vartheta})(\vartheta_{2(0)} + \vartheta_{2(22)}) + A] \\
&\quad + \bar{\vartheta}[(\bar{\mu}_1 g_{\mu\mu} + g_{\mu\vartheta})(\mu_{2(0)} + \mu_{2(22)}) + (g_{\vartheta\vartheta} + \bar{\mu}_1 g_{\mu\vartheta})(\vartheta_{2(0)} + \vartheta_{2(22)}) + B] \Big), \\
R &= \frac{1}{2} f_{\mu\mu\mu} \bar{\mu}_1^3 + \frac{3}{2} f_{\mu\mu\vartheta} \bar{\mu}_1^2 + \frac{3}{2} f_{\mu\vartheta\vartheta} \bar{\mu}_1 + \frac{1}{2} f_{\vartheta\vartheta\vartheta}, \\
S &= \frac{1}{2} g_{\mu\mu\mu} \bar{\mu}_1^3 + \frac{3}{2} g_{\mu\mu\vartheta} \bar{\mu}_1^2 + \frac{3}{2} g_{\mu\vartheta\vartheta} \bar{\mu}_1 + \frac{1}{2} g_{\vartheta\vartheta\vartheta}, \\
A &= f_{\mu\mu\mu} \bar{\mu}_1^3 + 3 f_{\mu\mu\vartheta} \bar{\mu}_1^2 + 3 f_{\mu\vartheta\vartheta} \bar{\mu}_1 + f_{\vartheta\vartheta\vartheta}, \\
B &= g_{\mu\mu\mu} \bar{\mu}_1^3 + 3 g_{\mu\mu\vartheta} \bar{\mu}_1^2 + 3 g_{\mu\vartheta\vartheta} \bar{\mu}_1 + g_{\vartheta\vartheta\vartheta}.
\end{aligned}$$

We will now employ amplitude stability analysis to examine the selection of Turing patterns. Let us suppose that every amplitude in eqn (10.49) can be divided as

$$W_j = \Psi_j \exp(i\Phi_j), \quad j = 1, 2, 3, \tag{10.51}$$

where phase angle is Φ_j, which corresponds to mode $\Psi_j = |W_j|, j = 1, 2, 3$. By using eqn (10.51) in eqn (10.49), we obtain

$$
\begin{aligned}
\Upsilon_0 \frac{\partial \Phi}{\partial t} &= -\ell \frac{{\Psi_1}^2 Psi_2^2 + {\Psi_2}^2{\Psi_3}^2 + {\Psi_3}^2{\Psi_1}^2}{\Psi_1\Psi_2\Psi_3} \sin\Phi, \\
\Upsilon_0 \frac{\partial \Psi_1}{\partial t} &= \rho\Psi_1 + \ell\Psi_2\Psi_3\cos\Phi - s_1{\Psi_1}^3 - s_2({\Psi_2}^2 + {\Psi_3}^2)\Psi_1, \\
\Upsilon_0 \frac{\partial \Psi_2}{\partial t} &= \rho\Psi_2 + \ell\Psi_3\Psi_1\cos\Phi - s_1{\Psi_2}^3 - s_2({\Psi_3}^2 + {\Psi_1}^2)\Psi_2, \\
\Upsilon_0 \frac{\partial \Psi_3}{\partial t} &= \rho\Psi_3 + \ell\Psi_1\Psi_2\cos\Phi - s_1{\Psi_3}^3 - s_2({\Psi_1}^2 + {\Psi_2}^2)\Psi_2,
\end{aligned}
\tag{10.52}
$$

where $\Phi = \Phi_1 + \Phi_2 + \Phi_3$. We can infer by the equation first in eqn (10.52) that when $\ell > 0$ and $\Phi = 0$ or $\ell < 0$ and $\Phi = \pi$, then the stability of the solution is confirmed. As a result, we get:

$$
\begin{aligned}
\Upsilon_0 \frac{\partial \Psi_1}{\partial t} &= \rho\Psi_1 + |\ell|\Psi_2\Psi_3 - s_1\Psi_1^3 - s_2(|\Psi_2|^2 + |\Psi_3|^2)\Psi_1, \\
\Upsilon_0 \frac{\partial \Psi_2}{\partial t} &= \rho\Psi_2 + |\ell|\Psi_1\Psi_3 - s_1\Psi_2^3 - s_2(|\Psi_1|^2 + |\Psi_3|^2)\Psi_2, \\
\Upsilon_0 \frac{\partial \Psi_3}{\partial t} &= \rho\Psi_3 + |\ell|\Psi_2\Psi_1 - s_1\Psi_3^3 - s_2(|\Psi_1|^2 + |\Psi_2|^2)\Psi_3.
\end{aligned}
\tag{10.53}
$$

By examining amplitude equations, it is possible to determine the many kinds of Turing patterns and also the stability of the dynamical model (10.19) that they represent. There are four different types of solutions to the dynamical system (10.52), which are covered in Table 10.1.

10.8 Numerical Simulations

In this section, numerical simulations are used to validate the analytical results from the previous sections. This would facilitate our explanation of the corresponding temporal dynamics and the spatiotemporal dynamics of systems (10.11) and (10.19). The following fixed parameters are used in the numerical simulations: $\sigma = 11$, $h = 0.8$, $d_\mu = 10$, and $d_\vartheta = 0.1$. As controlled parameters, α and K were also considered.

For a fixed value of $\alpha = 3$ and $K = 0.8$, the equilibrium points of system (10.11) are as follows: $(0, 0)$, $(0.8, 0)$, $(0.440724, 2.1772)$, and $(0.537535, 1.93991)$. We are interested in conducting research on the coexistence of both species in the system. The coexistence equilibrium has two points in this instance. -4.04283 and 0.0704042 are the computed eigenvalues at the coordinates $(0.537535, 1.93991)$. Thus, saddle system (10.11) at $(0.537535, 1.93991)$. In contrast, -1.18819 and -0.252284 are the obtained

Table 10.1 Stability and range of existence of the different kinds of patterns in eqn (10.52).

S.No	Type of solution	Formula	Range of existence	Range of stability
1.	Spot solution	$\Psi_1 = \Psi_2 = \Psi_3 = \frac{\lvert\ell\rvert \pm \sqrt{\ell^2+4(s_1+2s_2)\rho}}{2(s_1+2s_2)}$, $\Psi^+ = \frac{\lvert\ell\rvert + \sqrt{\ell^2+4(s_1+2s_2)\rho}}{2(s_1+2s_2)}$, $\Psi^- = \frac{\lvert\ell\rvert - \sqrt{\ell^2+4(s_1+2s_2)\rho}}{2(s_1+2s_2)}$.	$\rho > \rho_1 = \frac{-\ell^2}{4(s_1+2s_2)}$, $H_0: \ell > 0$, $H_\pi: \ell < 0$.	**Unstable**: (Ψ^-) always where $\rho_4 = \frac{2s_1+s_2}{(s_2-s_1)^2}\ell^2$ **Stable**: (Ψ^+), $\rho < \rho_4$
2.	Stripe solution	$\Psi_1 = \sqrt{\frac{\rho}{s_1}} \neq 0$, $\Psi_2 = \Psi_3 = 0$	$\rho > 0$	**Unstable**: $\rho < \rho_3$ where $\rho_3 = \frac{\ell^2 s_1}{(s_2-s_1)^2}$ **Stable**: $\rho > \rho_3$
3.	Stationary solution	$\Psi_1 = \Psi_2 = \Psi_3 = 0$	always	**Unstable** : $\rho > \rho_2 = 0$ **Stable**: $\rho < \rho_2 = 0$
4.	Mixed solution	$\Psi_1 = \frac{\lvert\ell\rvert}{s_2-s_1}$, $\Psi_2 = \Psi_3 = \sqrt{\frac{\rho - s_1\Psi_1{}^2}{s_1+s_2}}$.	$\rho > \rho_3$	**Unstable**: always

eigenvalues of the variational matrix at $(0.440724, 2.1772)$. Since the eigenvalues have a negative sign, they are real and unequal. Since system (10.11) is stable at this point, the equilibrium point $E'^* = (0.440724, 2.1772)$ can be used in numerical simulations.

The temporal behavior of system (10.11) is analyzed using Hopf bifurcation and Turing instability condition, which follow the stability of coexisting equilibrium point E'^*. As there wil not be enough prey to sustain the predator population, a low cooperation coefficient ($\alpha = 1, 2$) will cause it to go extinct (see Figure 10.1(a) and (b)). Predators cooperate when hunting; therefore, even in cases when $\alpha (= 3, 4)$ is significant, the predator survives (Figure 10.1 (c,d)). With α and K as the bifurcation parameters, the Hopf bifurcation diagrams of the prey and predator populations are shown in Figures 10.2 and 10.3. As can be shown in Figure 10.2, we now begin by changing the hunting cooperation α. The steady state of the predator extinction is unstable by the red curve and stable for conceivable values of the parameter α represented by the blue curve. A Hopf bifurcation (H) occurs for a small range of numeric values of $alpha = baralpha = 3.2093183$, meaning that the stable non-trivial steady state turns unstable. Moreover, at $K = barK = 0.83509383$, a Hopf bifurcation (H) occurs, as seen in Figure 10.2 and 10.3. Through the temporal analysis of system (10.11), we were able to discover the domain of cooperation coefficients (α) and (K) in which our model can survive.

10.8.1 Pattern selection

System eqn (10.5) for the spatiotemporal model will now be numerically simulated in the 2D spatial domain with initial condition and boundary conditions of zero flux. We select patterns in a two-dimensional space. The domain size is $[50 \times 50]$, with a time interval of $(\Delta t = 0.5)$ and a space interval of $(\Delta\mu = 0.5 = \Delta\vartheta)$. We explore the kinds of patterns that will emerge from the preceding study by adjusting the controlled parameter α for fixed parameters. We derive the Turing instability condition for several values of the diffusion coefficient (d_ϑ). Figure 10.4(b) plots the zone for the creation of spatial patterns that satisfy the Turing instability criterion versus the ratio of diffusion coefficients (d_ϑ). If the D dispersion relation is represented by $Re(\lambda)$ in the spatial model. The Re(λ) matched graph is shown in Figure 10.4(b) in relation to wave number (k). Since the wave number (k) fits in the interval (k_-, k_+), if $k \in (0.149879, 0.365857)$, then the real part of the eigenvalue Re$(\lambda) > 0$ is true. The remaining parameters for wave number k in the range $(0.149879, 0.365857)$, Re$(\lambda) > 0$, are mentioned in the caption of the picture. For more details, refer to Figure 10.4(b).

We obtain $s_1 = 0.6513$, $s_2 = 1.1356$, $\rho_1 = -0.0073$, $\rho_2 = 0$, $\rho_3 = 0.454$, and $\rho_4 = 1.2137$ with fixed $\alpha = 0.4$ and $s = 0.02$. The stability and unstability conditions as well as the conditions for the region of existence for the parameters that depend on ρ are listed in Table 10.1. Next, we have Table 10.2 the several values of ρ that correspond to various patterns for various diffusion coefficients picked from the Turing region (Figure 10.4).

Figure 10.5 shows the Turing patterns of prey and predator for various values of cooperation hunting coefficient α.

The 2D spatial patterns of system (10.19) with $\alpha = 2.88$ for the predator population and prey population are depicted in Figure 10.5(a) and (b). The spot pattern dominates the whole habitat in these figures. Prey are separated with small density, as seen in Figure 10.5(a), where black spots are scattered on a dark brown background. The predators are separated and have a high population density in Figure 10.5(b) with the black-yellow spots on the blue background. Prey density usually decreases with an increase in the hunting

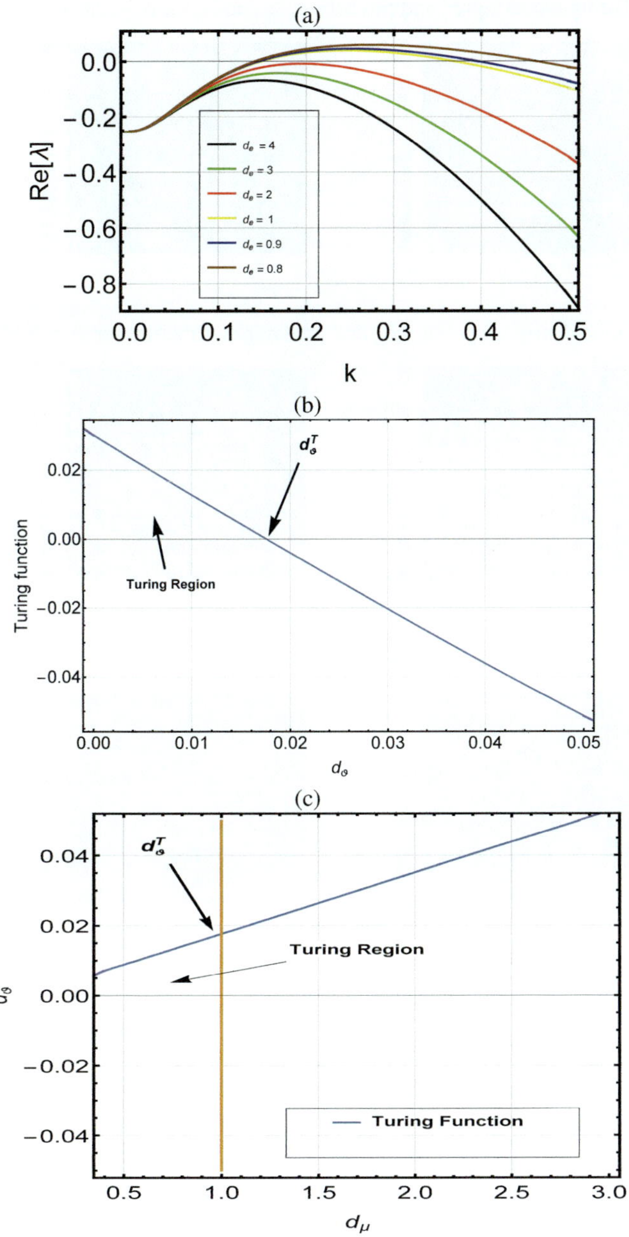

Figure 10.4 (a) Illustration of real part of λ against k. (b) Illustration of Turing function against d_ϑ. (c) Illustration of $(d_\mu - d_\vartheta)$ plane. The fixed parameters are $\mu = 0.440724, \vartheta = 2.1772, \sigma = 11, K = 0.8, h = 0.8, t = 100$, and $d_\mu = 1$

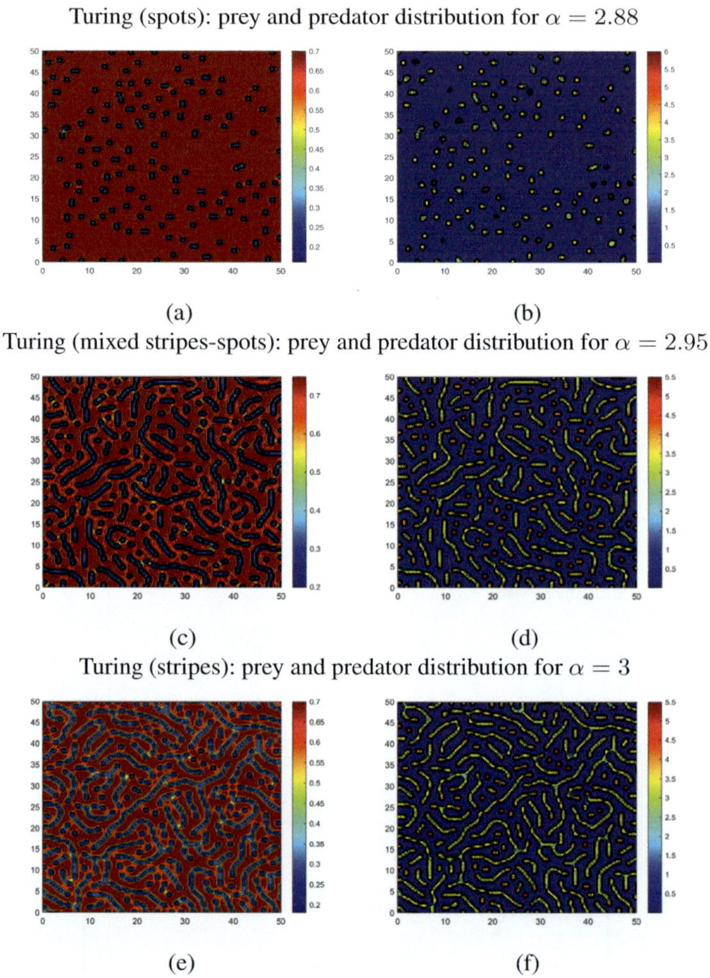

Figure 10.5 2D Turing patterns with parameter values $\sigma = 11$, $K = 0.8$, $h = 0.8$, $t = 100$, $d_\mu = 1$, and $d_\vartheta = 0.01$.

cooperation rate $(= 2.95, 3)$. High hunting cooperation rates enable the predator population to kill more prey, which lowers the prey population and, in turn, lowers the predator population since there are fewer prey animals available. Spot-stripe patterns develop in both the prey and predator populations as the cooperation rate is increased to 2.95 Figure 10.5(c) and (d). When α rises to 3, then the model shows a spot pattern decline and an emergence of stripes Figure 10.5(e) and (f). This results in a chain of spots, stripe-spots, and stripes being noticed as the value of hunting cooperation coefficient increases.

Table 10.2 For different values of d_ϑ, selection of patterns for eqn (10.19).

S.No	Value of d_ϑ	$\rho = \frac{d_\vartheta^T - d_\vartheta}{d_\vartheta^T}$	Region	Pattern	Figure
1.	0.010	0.4363	$\rho \in (\rho_2, \rho_3)$	Stripes	Figure 10.7(a) and (b)
2.	0.008	0.5490	$\rho \in (\rho_3, \rho_4)$	Stripes and spots	Figure 10.7(c) and (d)
3.	0.006	0.6617	$\rho \in (\rho_3, \rho_4)$	Stripes and spots	Figure 10.7(e) and (f)
4.	0.003	0.8308	$\rho \in (\rho_3, \rho_4)$	Stripes and spots	Figure 10.7(g) and (h)

We also obtained the Turing patterns by varying the value of the parameter K. For different values of K, system (10.19) gives three kinds of patterns. If $K = 0.78$, then the prey are separated and have a low population density, as can be shown in Figure 10.6(a), where the black dots are scattered on a dark brown backdrop.

The predators are separated and have a high population density in Figure 10.6(b), with the black-yellow spots on the blue background. With an increase in the basic reproduction number of the predator ($K = 0.78, 0.79$), the predator density also increases. Spot-stripe patterns develop in both the prey and predator populations as K is increased to 0.79 Figure 10.6(c) and (d). When K rises to 0.81, then the models shows a spot pattern decline and an emergence of stripes Figure 10.6(e) and (f). This results in a sequence of spots, spot-stripes, and stripes being noticed as the value of the parameter K increases. The Turing patterns were then also generated by adjusting the diffusion coefficient d_ϑ. Three types of patterns are produced by system (10.19) for various values of the d_ϑ (Figure 10.7). Stripes Turing patterns were seen in both the prey and predator populations if $d_\vartheta = 0.01$ (Figure 10.7(a) and (b)). As d_ϑ decreases to 0.008 (Figure 10.7(c) and (d)) and to 0.006 (Figure 10.7(e) and (f)), spot-stripe patterns emerge in both the prey and predator populations. The dynamics of the model exhibit spot patterns when d_ϑ drops to 0.003 (Figure 10.7(g) and (h)). As a result, as the value of d_ϑ drops, a chain of spots, stripe-spots, and stripes is observed. Predators can create cooperative hunting groups when there is a low rate of diffusion among them. Interspecific rivalry results from the reduced food availability among the predators caused by this constrained dispersion. Furthermore, there is a negative correlation between population density and diffusion rate. On the

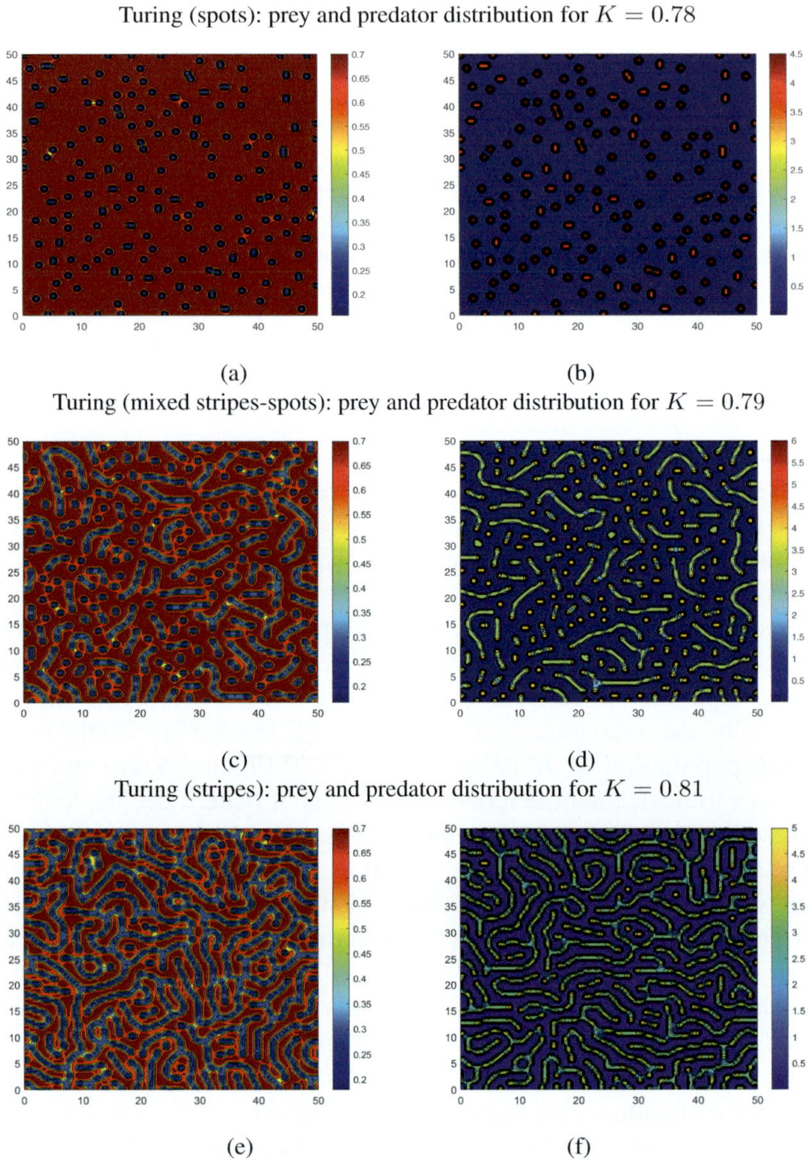

Figure 10.6 2D Turing patterns with parameter values $\sigma = 11$, $\alpha = 3$, $h = 0.8$, $t = 100$, $d_\mu = 1$, and $d_\vartheta = 0.01$.

other hand, the population density rises in tandem with the diffusion rate. The Turing pattern changes from a mixed pattern to a regular pattern as a result.

Turing (stripes): prey and predator distribution for $d_\vartheta = 0.01$

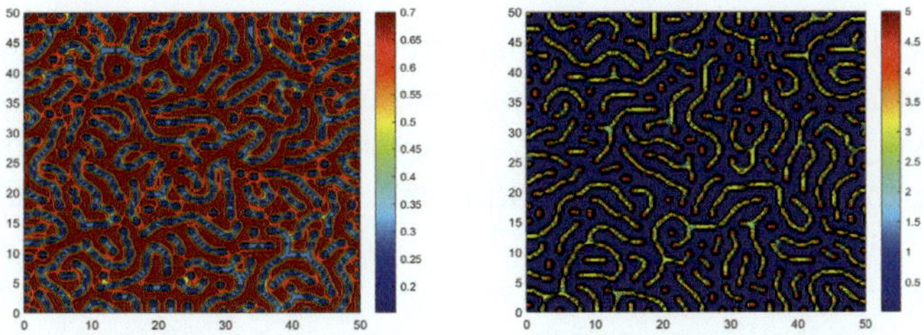

(a) (b)

Turing (mixed stripes-spots): prey and predator distribution for $d_\vartheta = 0.008$

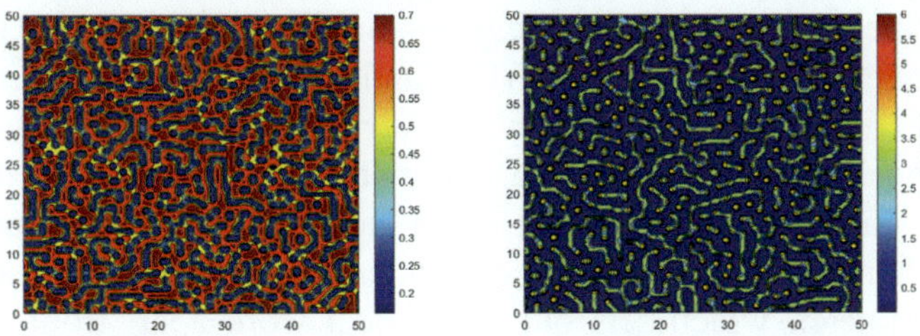

(c) (d)

Turing (mixed stripes-spots): prey and predator distribution for $d_\vartheta = 0.006$

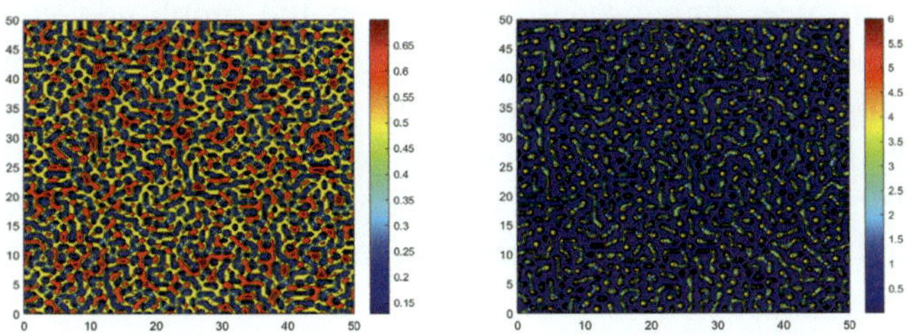

(e) (f)

Turing (mixed stripes-spots): prey and predator distribution for $d_\vartheta = 0.003$

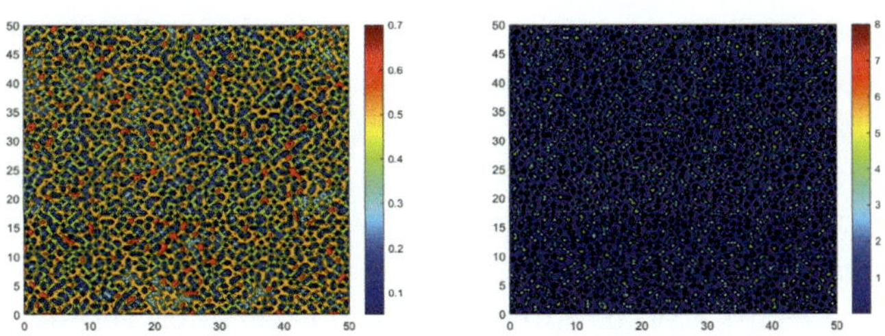

(g) (h)

Figure 10.7 2D Turing patterns with parameter values $\sigma = 11$, $\alpha = 3$, $h = 0.8$,$K = 0.8$, $t = 1000$, and $d_\mu = 1$.

10.9 Conclusion

Within biological systems, social animals frequently and broadly cooperate with one another. It is significant to note that all of these conclusions were drawn using a very basic predator–prey model, in which the benefits of cooperative hunting for the predator population are shown by simply adding the attack rate to a cooperation component. This study examines a predator–prey model by fusing cooperative predatory conduct with the herd behavior of the prey, together with a Holling type-II functional response in the prey–predator system.

The Turing pattern creation, spatiotemporal dynamics, and temporal dynamics have been used to discuss the prey–predator system. Then investigate the consequences of predators moving randomly through a geographical analysis of pattern development. Firstly, we verify that the current system is bounded and positive. The bifurcation parameter in our article is either K or the hunting cooperation coefficient α. In earlier studies, the bifurcation parameter is often either the diffusion coefficient or α. We select a separate bifurcation parameter, which sets our theoretical work apart from others'. The bifurcation parameter d_ϑ is produced by leveraging the predator's random movement to obtain the diffusion-driven instability condition. Weakly nonlinear analysis was then used to determine the amplitude equations' for the prey–predator system. The resulting amplitude equations analysis indicates that this system can generate three types of patterns. A pattern consisting of spots, spot-stripes, and stripes was found by the numerical simulations. We obtain these three types of patterns by changing the values of parameters α and K. We may obtain these three types of patterns by altering the value of parameter d_ϑ.

References

[1] Malthus, T.R., (1798). An Essay on the Principle of Population, J. Johnson in St. Paul's Churchyard, London.

[2] Vogels, M., Zoeckler, R., Stasiw, D. M. & Cerny, L. C. (1975). P. F. Verhulst's "notice sur la loi que la populations suit dans son accroissement" from correspondence mathematique et physique. Ghent, vol. X. *Journal of Biological Physics*, 3, 183-192.

[3] Waltman, P., (1983). Competition Models in Population Biology. *SIAM, Philadelphia*.

[4] Lotka, A. J. (1956). Elements of Mathematical Biology. *Dover Publications Inc., New-York*, 7(3), 208-208

[5] Volterra, V., Ancona, D. & Umberto, (1931). Vital Competition Between Species in the Marine Environment, *Orleans New printing company of Loiret*, 1-14.

[6] Anderson, C., Franks N. R. (2001). Teams in animal societies. *Behavioral Ecology*, 12(5), 534-540.

[7] Connor RC (2000) Group living in whales and dolphins. In: Mann J, Conner RC, Tyack PL, Whitehead H, editors. Cetacean societies: Field studies of dolphins and whales. Chicago: University of Chicago Press. pp. 99-218.

[8] Dugatkin, L. A. (1997). Cooperation among animals. *Oxford: Oxford University Press*, 240 pages.

[9] Packer, C. & Ruttan, L. (1988). The evolution of cooperative hunting. *The American Naturalist*, 132, 159-198.

[10] Boesch, C. & Boesch, H. (1989). Hunting behavior of wild chimpanzees in the TaÃŕ National Park. *American Journal of Physical Anthropology*, 78(4), 547-573.

[11] Bshary, R., Hohner, A., Ait-el-Djoudi, K. & Fricke, H. (2006). Inter specific Communicative and Coordinated Hunting between Groupers and Giant Moray Eels in the Red Sea. *Plos Biology*, 4(12), e431.

[12] Du, Y., Niu, B. & Wei, J. (2021). A predator−prey model with cooperative hunting in the predator and group defense in the prey. *Discrete and Continuous Dynamical Systems - B*, 1531-3492.

[13] An, Q. & Jiang, W. (2019). Spatiotemporal attractors generated by the Turing-Hopf bifurcation in a time-delayed reaction-diffusion system, *Discrete Continuous Dynamical System-B*, 24, 487-510.

[14] Holling, C. S. (1959). The components of predation as revealed by a study of small-mammal predation of the European pine sawfly. *The Canadian Entomologist*, 91(5), 293-320.

[15] Kazarinov, N. D. & Driessche, P. V. D. (1978). A model predator−prey system with functional response. *Bellman Prize in Mathematical Biosciences*, 39, 125-134.

[16] Song Y., Peng, Y. & Zhang, T. (2021). The spatially inhomogeneous Hopf bifurcation induced by memory delay in a memory-based diffusion system, *Journal of Differential Equations*, 300, 597-624.

[17] Song, Y., Shi, J., & Wang, H. (2022). Spatiotemporal dynamics of a diffusive consumer-resource model with explicit spatial memory. *Studies in Applied Mathematics*, 148, 373-395.

[18] Braza, P. A. (2012). Predator-prey dynamics with square root functional responses. *Nonlinear Analysis: Real World Applications*, 13(4), 1837-1843.

[19] Bulai, I. M., & Venturino, E. (2017). Shape effects on herd behavior in ecological interacting population models, *Mathematics and Computers in Simulation*, 141, 40-55.

[20] Djilali, S. (2019). Impact of prey herd shape on the predator−prey interaction. *Chaos Solitons & Fractals*, 120 , 139-148.

[21] Kooi, B. W., & Venturino, E. (2016). Ecoepidemic predator−prey model with feeding satiation, prey herd behavior and abandoned infected prey, *Mathematical Biosciences*, 274, 58-72.

[22] Raw, S. N., Mishra, P., Kumar, R., & Thakur, S. (2017). Complex behavior of prey−predator system exhibiting group defense: A mathematical modeling study. *Chaos, Solitons & Fractals*, 100(C), 74-90.

[23] Xu, C., Yuan, S., & Zhang, T. (2016). Global dynamics of a predator−prey model with defense mechanism for prey. *Applied Mathematics Letters*, 62, 42-48.

[24] Freedman, H. I. & Wolkowicz, G. S. K. (1986). Predator-prey systems with group defense: The paradox of enrichment revisited, *Bulletin of Mathematical Biology* 48(5-6), 493-508.

[25] Ali, S. J., Arifin, N. M., Naji R. K., Ismail, F., Bachok, N. (2016). Analysis of Ecological Model with Holling Type IV Functional Response. *International Journal of Pure and Applied Mathematics*, 106(1), 317 - 331.

[26] Gakkhar, S., & Naji, R.K. (2002). Chaos in three-species ratio dependent food chain.*Chaos, Solitons & Fractals*, 14(5), 771-778.

[27] Gakkhar, S., & Singh, A. (2012). Control of chaos du to additional predator in Hastings-powell food chain model. *Journal of Mathematical Analysis and Applications*, 385, 423-438.

[28] Ajraldi, V., Pittavino, M., & Venturino, E. (2011). Modeling herd behavior in population systems, *Nonlinear Analysis: Real World Applications*, 12(4), 2319-2338.

[29] Venturino, E. & Petrovskii, S. (2013). Spatiotemporal behavior of a prey−predator system with a group defense for prey. *Ecological Complexity*, 14, 37-47.

[30] Belvisi, S., & Venturino, E. (2013). An ecoepidemic model with diseased predators and prey group defense. *Simulation Modelling Practice and Theory*, 34, 144-155.

[31] Yuan, S., Xu, C., & Zhang, T. (2013). Spatial dynamics in a predator−prey model with herd behavior. *Chaos*, 23(3), Article ID 033102.

[32] Brentnall, S. J., Richards, K. J., Brindley, J. & Murphy,E. (2003). Plankton patchiness and its effect on larger-scale productivity. *Journal of Plankton Research*, 25(2), 121-140.

[33] Fulton, E. A., Smith, A. D. M., & Johnson, C. R. (2003). Mortality and predation in ecosystem models: is it important how these are expressed?. *Ecological Modelling*,169(1), 157-178.

[34] Xu, C., & Yuan, S. (2014). Stability and Bifurcation Analysis of Differential Equations and its Applications. *Hindawi Publishing Corporation*, 2015(343528), 1 page.

[35] Singh, T. and Banerjee, S. (2019). Spatiotemporal model of a predator-prey 535 system with herd behavior and quadratic mortality. International Journal of Bifurcation and Chaos, 29(04), 1950049.

[36] Shivam, Singh, K., Kumar, M., Dubey, R., & Singh, T. (2022). Untangling role of cooperative hunting among predators and herd behavior in prey with a dynamical systems approach. Chaos, Solitons & Fractals, 162, 112420.

[37] Shi, X., Cui, J. & Zhou X. (2011). Stability and Hopf bifurcation analysis of an eco-epidemic model with a stage structure.

11

Aryabhata's Contribution to Advancement in Arithmetic

Gajendra Pratap Singh and Chosching Lazes

School of Computational and Integrative Sciences, Jawaharlal Nehru University, India
E-mail: gajendra@jnu.ac.in; clazes1297@gmail.com

Abstract

Aryabhata I was popularly known as the father of Indian mathematics. At 23, he wrote a treatise, Aryabhatiya, in Sanskrit, covering diverse concepts, including fundamental astronomy, mathematics, time calculation, and spherical geometry [3]. This chapter delves into the life story of Aryabhata and understanding of mathematical concepts included in "Aryabhatiya," such as the discovery of zero, and methods for solving complex mathematical concepts in a clear and accessible manner. The study highlights Aryabhata's mathematical knowledge as the foundation for further advancements in many mathematical concepts and their profound impact on the field. This study also showcases ancient mathematics and the intelligence of Indians, which enriches the concept of ancient mathematics, inspires future research, and continuously inspires students and mathematicians worldwide.

Keywords: Zero, square, square root, cube, cube root, method of inversion

11.1 Introduction

The great Indian mathematician and astronomer Aryabhata, also famous as Aryabhatta, was born in 476 CE. He lived in the Gupta period of Ancient India [2, 3]. "Bhata" means warrior and brave soldier, and "Bhatta" is a name used for Brahmin and philosopher. So, Aryabhatta with "tt" is more suitable for his characters. Many historians believe that he was born in

Kusumapura. One name mentioned was Patliputra, known as Patna, Bihar in India. From many facts, it is disclosed that he followed the Vedantic religion of Hinduism. He belongs to the Brahmin family but is not particular about family members. He was married and had a son named Devarajan, who later published an astronomical work. He completed his school and advanced studies in mathematics and astronomy at Kusumapura, where it also mentions that he was head of the institute (Kalupa), which is assumed as Nalanda University. Nalanda University was once famous for studying Brahmanical, Buddhist, philosophical, science, and arts; it was an essential learning center for Indian and international students. He learned six auxiliaries of the Veda, arts, and epistemology from his teacher and mastered probably; he knows three languages: Prakrta, Apabhramsa, and Sanskrit. He used to be a great, popular teacher by profession. Aryabhata was one of the pioneers in many fields, and everyone has appreciated him. In the sun temple Telangana, Bihar, he constructs an observatory. He writes books on mathematics and astronomy, but some need to be recovered. Aryabhatiya, the Arya-Siddhanta, and Surya Siddhanta-Prakasa were accessible books. By the age of 73, which is around 550 CE, he died. However, more resources are needed to assume when, how, and where he died [3].

11.2 Aryabhatiya: Treatise of Aryabhata

The book "Aryabhatiya," a treatise of Aryabhata, written at the age of 23 in Sanskrit, calculated and assumed his birth year from this [8]. In the book Aryabhatiya, there are four main sections as follows:

Gitikapada (basic of astronomy and sine)
Ganipada (mathematics)
Kalakriyapada (the calculation of time)
Golapada (the sphere)

He wrote particularly on mathematics concepts like discoveries of place value system and zero, decimal system, numerical approximation techniques, formulae for the area of a right triangle, right pyramid, technique to calculate cube and cube root, square and square root, solution to linear equation (as he is also known as father of Indian algebra), and trigonometric function such as sine. He did not directly prove the value of pie(π), but he gave the circumference of the earth and diameter of the planet later when they calculated the value of pie was 3.1416, which is correct up to two decimals. He also worked on the sum of power series [5, 10].

Moreover, he came up with many discoveries, like the earth's shape is a sphere that revolves around the sun and rotates on its axis, where he calculated the length of the year as 365 days, 6 hours, 12 minutes, and 30 seconds, which is approximately the same. He believes that the orbit of the planets is elliptical, and the moonlight and planet brightness are the reflection of sunlight. Furthermore, he explained solar and moon eclipse and their causes [6, 8].

In spherical astronomy, he applied plane geometry to address spherical geometry by projecting points and lines from the sphere's surface onto appropriate planes [8].

11.3 Seeding of Arithmetic by Aryabhata

11.3.1 Using the Sanskrit alphabet for enumeration

क	ka	1	ख	kha	2	ग	ga	3	घ	gha	4	ङ	ṅga	5
च	ch	6	छ	cha	7	ज	ja	8	झ	jha	9	ञ	ñya	10
ट	ta	11	ठ	tha	12	ड	da	13	ढ	dha	14	ण	na	15
त	ta	16	थ	tha	17	द	da	18	ध	dha	19	न	na	20
प	pa	21	फ	pha	22	ब	ba	23	भ	bha	24	म	ma	25
य	ya	30	र	ra	40	ल	la	50	व	va	60	श	sha	70
ष	sha	80	स	sa	90	ह	ha	100						

11.3.2 Sanskrit

The first ten national places are written as unit places as *Eka*, ten places as *dasa*, hundred places as *sata* is the hundred place, thousand places as *Sahasra*, ten thousand places as *ayuta*, hundred thousand places as *niyuta*, million places as *prayuta*, ten thousand million places as *koti*, hundred million places as *arbuda*, and thousand million places as *vrnda* respectively [3, 8].

These national places are denoted by writing zero as follows:

0 0 0 0 0 0 0 0 0 0

Hence, Aryabhata did not directly discover the zero but for the empty position he used a dot or small circle, which he later discovered as zero.

11.3.3 Square

"The product of two equal quantities." He did not explain how to find a square.

11.3.4 Square root

He explained the method to find a square root. "Subtract the largest possible square from the rightmost odd position and then write down the square root of the subtracted number in the square root line. Continuously divide the even positions (placed on the right) by twice the square root. Next, subtract the square of the quotient from the odd locations that are farthest to the right. Write the quotient in the next position, the number previously noted in the square root line. This will represent the square root. If more digits are on the right, continue this procedure" [1, 2, 3].

Example: 55696

Step 1: Odd and even should be marked from right to left.

$$5^{odd}5^{even}6^{odd}9^{even}6^{odd}$$

Step 2: Find the root nearest to the last odd place

$$r_1 = 2 \text{ is the root nearest to the last odd place, 5}$$

Step 3: Square the root and subtract to the last odd place

$$2^2 = 4$$

Step 4: Next, even place digit will be written down with the remainder and divided by twice the previous root. Quotient will be the next root and placed in a row after the previous one.

$$2*2 = 4 \quad \text{quotient} = 3 \text{ is the next root.}$$

Step 5: Continue the third and fourth steps till the remainder becomes zero. Complete step by an odd place should be subtracted with the square of the last root, and even place should be divided with twice every root.

$$3^2 = 9 \qquad 23^*2 = 46$$

Solve the previous step as shown below:

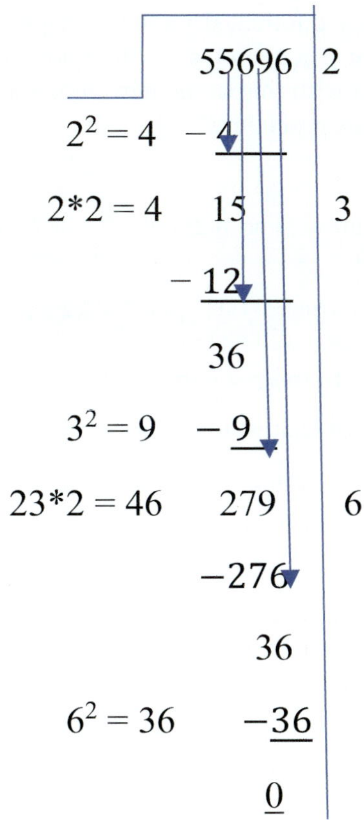

Hence, 236 is the square root of 55696.

11.3.5 Cube

"The product of three equal quantities." He gave no further explanation.

11.3.6 Cube root

To find the cube root of a number "subtract the most significant possible cube from the last cube positions and then write down the cube root of the subtracted number in the cube root line, divide the second noncubic place by three times the square of the obtained cube root, then subtract the cube of the quotient from the first noncubic place, which is located to the right of second noncubic place, multiplied by three times the previous cube root. Then subtract the cube of the quotient from the cube place, which is located to the right of the first cube place, and write the quotient on the line of the cube root, and then proceed with this as the new cube root; if there are more digits on the right, repeat this procedure" [3].

Example: 1225043

Step 1: Cubic place, non-cubic1, and non-cubic2 place, respectively, should be marked from right to left.

$$1^{\text{cubic}}2^{\text{non-cubic2}}2^{\text{non-cubic1}}5^{\text{cubic}}0^{\text{non-cubic2}}4^{\text{non-cubic1}}3^{\text{cubic}}$$

Step 2: Find the root nearest to the last cubic place

Quotient is = 1 is the root nearest to the last cubic place 1

Step 3: Cubing the root and subtract to the last cubic place

$$1^3 = 1$$

Step 4: Next non-cubic1 place digit will get down with remainder and divide with thrice the previous root square. Quotient will be the next root and placed in a row after the previous one.

Divisor is $3*1^2 = 3$ quotient is = 0 is the next root.

Step 5: Next, the non-cubic2 place digit will get down with the remainder and subtract with the product of thrice the previous root and square of next root.

Subtract from $3*1*0^2 = 0$

Step 6: The next cubic place digit will get down with the remainder and subtract the cube of the next or second root.

Subtract the $0^3 = 0$

Step 7: Next non-cubic1 place digit will get down with the remainder and divide with thrice the square of both root and write quotient below the previous quotient.

Divisor is $3*10^2 = 300$ quotient is 7

Step 8: Next, get down the non-cubic2 place with the remainder and subtract the product of thrice the previous first two roots and the square of the third root.

Subtract the $3*10*7^2 = 1470$

Step 9: Next, get down the cubic place with the remainder and subtract the cubic of the third roots. Continue Step 3–5 until the remainder becomes zero.

$$7^3 = 343$$

Solve the previous cube root as below and observe it before repeating each step.

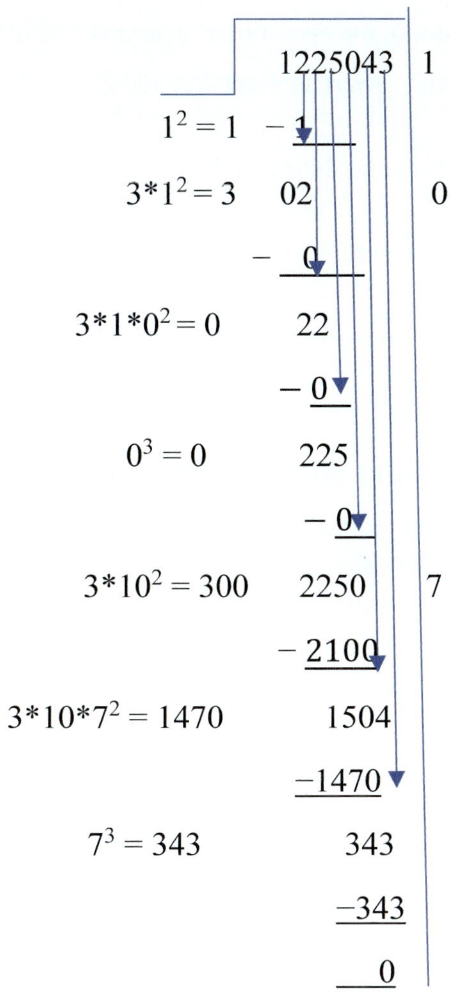

Hence, 107 is the cube root of number 1225043.

11.3.7 Method of inversion

"In the method of inversion, the multiplier changes to the divisor, and the divisor changes to the multiplier, and addition changes to subtraction, and where subtraction changes to the addition" [2, 3].

Example: If a number is divided by 2, added by 3, multiplied by 5, and then subtracted by 2, it leads to the result 3. What is the initial number?

Solution: By applying the method of inversion.

Starting from the last digit 3, the operation is also inverted in reverse order.

Step 1: From the last result, invert subtract 2 to add 2
$3 + 2 = 5$

Step 2: Invert multiplied by 2 to divide by 5
$5 \div 5 = 1$

Step 3: Invert added by 3 to subtract by 3
$1 - 3 = -2$

Step 4: Invert divided by 2 to multiply by 2
$-2 \times 2 = -4$
The result is -4.

11.4 Significance of the Work in Modern Technology

The significance of square, square roots, cube, cube roots, and method of inversion in modern technology as:

- Algorithms and computing: Both directly and indirectly, practical methods for computing powers and roots are frequently used in algorithms for numerical analysis, optimization, and simulation.
- Data encryption and cryptography: Modular arithmetic, exponential, and inversion methods are frequently used in cryptography algorithms.
- Scientific computing and simulation: To solve equations and analyze data, numerical models and simulations of physical systems frequently call for computations involving squares, square roots, cubes, and cube roots.

11.5 Some of the Honors in the Name of Aryabhata

In today's science and technology world, the rest of the work has become much easier because of his deed. To honor him, the Indian government named the first satellite Aryabhata and the lunar crater Aryabhata. Also, to honor him, the state government of Bihar built Aryabhata Knowledge University (A.K.U.) in Patna. The Astronautical Society of India (A.S.I.) gives the Aryabhata Award to people whose lifetime work in the field of astronautics and aerospace technology in India [4].

11.6 Conclusion

Aryabhata, an ancient Indian mathematician and astronomer, contributes to many fields of mathematics and astronomy, like seeding for future advanced work. Through his observations, curiosity, methodologies, and critical thinking, he has discovered a lot of knowledge about the world. His pioneer's work, such as Aryabhatiya, provided an understanding of mathematics and findings on astronomy through innovative techniques. His contributions to trigonometry, arithmetic, algebra, geometry knowledge, measurement approach, and astronomy have lasted through the ages, influencing many students and scholars and shaping mathematics further [8].

As we reflect on his work and discoveries, we know his role in shaping and understanding mathematics, astronomy, and the world around us. Through Aryabhata's brilliance and innovation, he left an enduring legacy that continuously enriched and inspired the world of mathematics and astronomy. At present, there is a need to connect the Aryabhata for the betterment of science and society to study the intermediate states with graph theory, communicable diseases [13, 17], computational biology, bioinformatics, computer science, and Petri nets [14], machine learning, AI [15], etc. It will be a significant nation-building enhancement if we successfully reduce the computational running time of any real-time-based algorithm.

Acknowledgments

Gajendra Pratap Singh acknowledges the support from the Department of Science and Technology (DST)-Science and Engineering Research Board (SERB) project (MATRICS ID: *MTR/2021/000378*), M.Sc. consumable grant, DBT BIC and the Department of Biotechnology, Ministry of Science & Technology, Govt. of India (Project ID: BT/PR40251/BITS/137/11/2021), and ICMR Project ID 2021-15450 for providing the research facility. The authors would also like to thank the reviewers and readers for their valuable feedback.

References

[1] Parajuli, Krishna Kanta. "A Comparative Study on Three Pioneer Methods for the Square Roots." *Nepal Journal of Mathematical Sciences* 2, no. 2 (2021): 89-96.

[2] Clark, Walter Eugene. "The aryabhatiya of aryabhata: An ancient indian work on mathematics and astronomy." (1930).

[3] Smith, David Eugene. "The Aryabhatiya of Aryabhata." (1930): 396-398.

[4] Bag, A. K. "Aryabhata I's achivements in ancient Indian mathematics & Astronomy and their relevance in the Twenty-first century." *Journal of the Asiatic Society of Bengal* 47, no. 1 (2005): 14-26.

[5] Gongol, William J. "The aryabhatiya: Foundations of indian mathematics." (2003).

[6] Raju, C. K. "Aryabhata Dalit, his philosophy of ganita, and its contemporary applications1." In *Theory and Praxis*, pp. 139-152. Routledge India, 2020.

[7] Fleet, John Faithfull. "V. Aryabhata's System of expressing Numbers." *Journal of the Royal Asiatic Society* 43, no. 1 (1911): 109-126.

[8] Ansari, S. M. "Aryabhata I, his life and his contributions." (1977).

[9] Yano, Michio. "Aryabhata's Method of Expressing Numbers." *Journal of Indian and Buddhist Studies (Indogaku Bukkyogaku Kenkyu)* 19, no. 1 (1970): 438-434.

[10] Kern, Hendrik, ed. *The Aryabhatiya*. Brill Archive, 1874.

[11] Singh, Gajendra Pratap, et al. "Modeling the mechanism pathways of first line drug in Tuberculosis using Petri nets." *International Journal of System Assurance Engineering and Management* 11 (2020): 313-324.

[12] Singh, Gajendra Pratap, et al. "A Graph-Theoretic Analysis on Functional EEG Network in Igraph R." *Decision Analytics Applications in Industry* (2020): 541-555.

[13] Singh, Gajendra Pratap, and Agraj Gupta. "A Petri net analysis to study the effects of diabetes on cardiovascular diseases." *2019 6th International Conference on Computing for Sustainable Global Development (INDIACom)*. IEEE, 2019.

[14] Singh, Gajendra Pratap, and Sujit Kumar Singh. "Petri net recommender system for generating of perfect binary tree." *International Journal of Knowledge and Systems Science (IJKSS)* 10.2 (2019): 1-12.

[15] Singh, Gajendra Pratap, Sangita Kansal, and Mukti Acharya. "Embedding an arbitrary 1-safe Petri net into a boolean Petri net." *International Journal of Computer Applications* 70.6 (2013).

[16] Kansal, Sangita, Gajendra Pratap Singh, and Mukti Acharya. "On Petri nets generating all the binary n-vectors." *Scientiae Mathematicae Japonicae* 71.2 (2010): 209-216.

[17] Singh, Gajendra Pratap, Aparajita Borah, and Sangram Ray. "A review paper on corona product of graphs." *Advances and Applications in Mathematical Sciences* 19.10 (2020): 1047-1054.

12

A Brief Account of Ancient Indian Astronomy and Mathematics

D. S. Hooda[1] and Gajendera Pratap Singh[2,*]

[1]Kurukshetra University, India
[2]School of Computational and Integrative Sciences, Jawaharlal Nehru University, India
E-mail: ds_hooda@rediffmail.com; gajendra@jnu.ac.in
*Corresponding author

Abstract

In the present communication, a brief history of ancient Indian astronomy covering pre-Siddhanta and the Siddhanta Jyotish periods is given. In literature, there are five Siddhanta books written by Sages; some other books on astronomy written by outstanding astronomers are also discussed. Historical development of ancient Indian mathematics is described with examples. The development of decimal system and numerals with zero symbols are discussed in detail. In the end, some concluding remarks about ancient Hindu mathematics and astronomy are also given with references.

Keywords: Ancient Indian Astronomy, Vedanga Jyotish, Surya Siddhanta, Sulva Sutra, Decismal System, Lunar Calendar

12.1 Introduction

Many authors have opined that life in ancient India was dominated by religion. Mathematics was created for designing yagna kunds and altars for various types of havans [1]. Astronomy was developed to regulate religious rituals. That was why more mathematics was created to meet the needs of astronomy. Most of Hindu astronomy was given in religious books, and most of Hindu mathematics was also given in books on Hindu astronomy [2, 3].

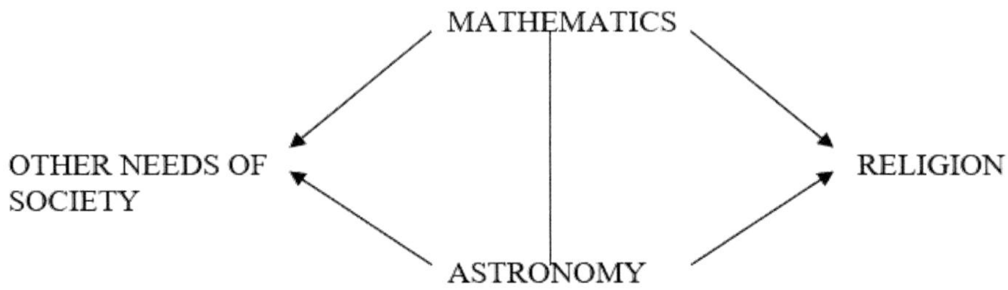

Figure 12.1 Relation between mathematics and astronomy.

The close relationship between mathematics and astronomy has continued in modern times also, though mathematics has many users other than astronomers (and religion does not need mathematics very much now). Moreover, modern astronomy also depends on many other disciplines like physics, chemistry, biology, etc., and these were not developed to the same extent as mathematics in ancient India.

12.2 Brief History of Ancient Indian Astronomy

For convenience, this history is divided into two periods: (a) the pre-Siddhanta period and (b) the Siddhanta Jyotish period [1, 2].

(a) The pre-Siddhanta period:

The pre-Siddhanta period is further divided into four parts [3]:

(i) Pre-Vedic astronomy: Though there is no direct reference to astronomy in the antiquities found in the excavations of Mohenjo-daro and Harappa, it is believed that such a highly developed civilization would have required a calendar based on astronomy.

(ii) Vedic period astronomy:

- There are many astronomical references in the *vedas*, particularly the *Yajurveda*, which show that a system of astronomy based on practical observations of the moon and stars existed to provide the calendar to enable the priests to perform *yagnas*, which had to be performed in different parts of the year and at different points of the day.
- A day was reckoned from sunrise to sunset and the night was reckoned from sunset to sunrise. Each was divided into 15 *mahurats*. The *mahurats* were named differently for each day and night of the dark and bright fortnights of a lunar month.

- Besides these, there were three (*Purvahna*, *Madhyahna*, and *Sayahna*) and five (*Pratah*, *Sangava*, *Madhyama*, *Aparahna*, and *Sayahna*) divisions of a day for a rough division of time. The lunar calendar used was based on lunar months, and all 30 days of the lunar month were given specific names. There were two groups of Vedic people; one reckoned a month from full moon to full moon, and the other from new moon to new moon.
- A year consisted of 12 months (Madhu, Madhava, Shukra, Shuchi, Nabha, Nabhasya, Isha, Urja, Sahas, Sahasya, Tapas, and Tapasaya), spread over six seasons (Vasanta, Grishma, Varsha, Sharad, Hemanta, and Shishira). Later, with the discovery of Nakshatras, the names of months were changed to their present names: Chaitra, Vaishakh, etc. The solar year was also started to be used along with the lunar month, and a 13th month, Amhaspati, was introduced to regulate calculations.

There were two Aryans in a year, namely Uttarayana and Dakshinayana, marking the period of northward and southward motions of the sun. A Vishwa day (an equinox day) is mentioned between the two Aryans. A yuga is also mentioned, but it is meant only for a period of five years in the Vedic period. Within the belt along which the Moon moves around the path of the Sun (ecliptic), wherein its maximum celestial longitude is five degrees north or south, a Nakshatra system was supposed to consist of 27 Nakshatras, all of which were named in Vedic times were slightly different from their present names. Among the planets, Brahspati (Jupiter), Shukra (Venus), and Rahu are mentioned, but it is not easy to see how those who constantly watched the Nakshatras could have left the other planets unnoticed. Comets and meteors (Dhumketu and Ulkas) are clearly mentioned. There is no mention of a seven-day week or zodiac signs (rashis) in the Vedas.

(iii) Vedanga Jyotish period of astronomy:

For every *Veda*, rules were given for calculating astronomical results in a text called Vedanga Jyotish supposed to be written by the unknown writer Leghadha. The period might be about the thirteenth-century BC. A period of five years was called *Yuga*, and the names of the five years were *Samvastar*, *Parivastar*, *Idavastar*, *Anuvastar*, and *Idvastar*. A solar year was supposed to consist of 366 days. A Yuga was supposed to consist of 1830 civil days, 62 (synodic) months of the moon, 30 omitted tithes, 2 intercalary months, and 5 mean revolutions of the sun. Vedanga Jyotish dealt only with rough calculations of solar and lunar elements. There were no calculations for planets.

(iv) Mahabharata period of astronomy:

During this period, Indians had knowledge of five planets; they observed them in the framework of Nakshatras belt and in conjunction with one another. They knew about the retrograde motions of planets and they attached astrological significance to all these phenomena. They had a time scale of a Kalpayuga of 4.32×109 years and of a Mahayuga of 4.32×106 years consisting of Kalyuga of 432,000 years, Dwapuryuga of 864,000 years, Treta Yuga of 1,296,000 years, and Satyuga of 1,728,000 years.

(b) Siddhanta Jyotish period of Indian astronomy:

Siddhanta Jyotish means a system that gives rules for calculating time, motions of planets, etc., from the beginning of the universe (Sreshtiad) to the end (pralya). A book is called Karana if it gives rules for calculating astronomical events from a fixed epoch other than the beginning of the *yuga of Kalpa*. The period in which these books were written is called the Siddhanta Jyotish period.

The following are the five Siddhanta books that are supposed to be written either by immortal beings or by sages:

(i) Paithamaha or Brahma Siddhanta
(ii) Vasishtha Siddhanta
(iii) Romaka Siddhanta
(iv) Paulisa Siddhanta

(iv) Surya Siddhanta:

The most respected and scientific of these is the Surya Siddhanta, and its calculations are based on recorded coordinates of Nakshatras. It is believed to have been written in about AD 280.

It is claimed that the third and fourth of the Siddhanta books may have foreign influence since these have, in some places, greater resemblance to Greek astronomy than to Indian astronomy. However, the most accurate of these, viz. the Surya Siddhanta is certainly Indian. All these five Siddhantas were edited by the famous astronomer Varaha Mihira in his famous work Pancha Siddhantika.

The following are the other books on astronomy that were written later by some outstanding mathematicians and astronomers:

1. Aryabhatiya: It was written by Aryabhata 1 (421 Shaka), whose fifteen hundredth anniversary we celebrated in 1976 and after whom we named

our first satellite. This was the first book on astronomy that is supposed to be written by a non-divine being. The book also contains chapters on arithmetic, algebra, and trigonometry and gives the first tables for trigonometric functions. Details about this book are given in Chapter III "The Aryabhatiya."

2. Brahma Sphuta Siddhanta (revised system of Brahma) by Brahmagupta (550 Shaka): This book gives accurate rules for finding positions of planets and corrects errors in earlier works. Bhaskarachrya considered Brahmagupta to be the "Ganit Chakra Churamani" or first gem among the circle of mathematicians.
3. Baswati Karan by Shatanand (1021 Shaka): This book uses centimal multiplication and division system.
4. Siddhanta Shiromani by Bhaskaracharya (1036 Shaka): This book is the most comprehensive one and gives a complete account of Indian astronomy till that time. It also discusses the obliquity of the ecliptic and its effect on the equation of time. This contains four parts, namely Lilavati, Bijaganita, Goladhaya, and Grahaganita.
5. Grahalagyam by Ganesh (1442 Shaka): This book is useful for those who have only limited knowledge of mathematics.
6. Siddhanta – Tattave Viveka by Kakalaka Bhatt (1530 Shaka): It is a good book on mathematical astronomy.

Some books written in Kerala:

1. Deva Keralam, Sukra keralam, of about the seventh-century AD.
2. Sarkaranarayaneeyam, by Sarkaranarayana, of the ninth-century AD. This book refers to an astronomical observatory and its instruments established by a contemporary king, Ravi Verma.
3. Venwaroham by Mahadevan, which discusses trigonometry and planetary motions.
4. Karan Paddhati written in AD 1431.
5. Sadratnamala by Sakara Vaiman, written in about AD 1530.
6. Tantra Samgraham by Nilkantha, written in about AD 1500.
7. Yukti Bhasa by Brahmadatta, written in AD 1639.

The last five books are in verse form and give infinite series expressions for sin x, cos x, and tan x about two and half centuries earlier than in Europe. The series known as Gregory's series for tan x was also known in India more than two centuries earlier than in Europe.

The use of instruments is essential to the efficacy of astronomy because it is a practical subject. This was something that the astronomers of ancient

India recognized, and they created tools that were appropriate for measuring astronomy. A chapter on the design and operation of such equipment may be found in nearly every astronomy book.

The earliest instrument available in Vedic times was *Turya*, consisting of a quadrant circle with a movable tube through the center and resting on the arc of the quadrant, which is graduated in degree and minutes. Through the tube, the heavenly bodies were observed, and the angle made by the tube with the vertical gave the zenith distance called *Natansk* and its complement *unantash* or altitude. During the Vedanga Jyotish period, a water clock was developed.

Chapter XIII of Surya Siddhanta discusses the construction of an auxiliary sphere, a *Sankuyantra or gnomon*, the *Cakrayantra*, *Mayur yantra*, *vanayantra*, etc. Brahmagupta discusses, in Brahma Sphuta Siddhanta, 17 instruments by which observations could be made. Bhaskaracharya gave the description of an additional instrument called Phalak yantra. Padmanabhan devised another instrument called Dhruvmrey Yantra in the fourteenth century. Yantra Cintamani, written by Chakardan, was a separate book on instruments.

An observatory appears to have been constructed at Ujjain or Avanti; it may have been at the same place where the present observatory stands. Another observatory, Man Mandir, was constructed by Raja Man Singh at Varanasi, and a number of instruments were added to it by Raja Jai Singh, who also built the observatory at Jaipur as well as the city of Jaipur around it. He described all instruments except rasivlaya, for which an effort was made by Garret in 1907 to reconstruct it. However, in the process, he changed it completely. Raja Jai Singh also built the observatory at Delhi

12.3 Some Comments on Ancient Indian Astronomy

We start by giving the reference by O. Neugebauer's book "A History of Ancient Mathematical Astronomy". A scholarly work comprising 3 volumes and more than 1200 pages, published in 1976.

The study of Hindu astronomy is still in its infancy, despite the groundbreaking work done by H. T. Colebrook (1765–1837), G. Thibault (1848–1914), and others. There exists a vast amount of unexplored manuscript material both in India and western countries. Would it suffice to note that American libraries house hundreds of planetary tables readily accessible for study? However, this material has just been the subject of a preliminary research, which has revealed many parameters for lunar planetary tables.

The planetary tables are extensive and employ methods not yet encountered in western materials, and the fundamental concept revolves around computing planetary position for an entire year on a function of Aries. The exact period of origin for these methods is uncertain; however, the extant texts indicate a possible origin around the fourteen-century AD.

This statement clearly establishes the need for a massive research effort. It is also a testimonial to the originality of Indian work on astronomy.

There is a feeling that there has been some Greek influence on Indian astronomy, and one of the two Siddhantas shows this influence. However, the Surya Siddhanta, which is purely Indian, is far superior to them, and this was due entirely to the great mastery of ancient Indians over arithmetic and algebra, while the Greeks were superior in geometry.

The independence of the two developments is testified by the fact that the parameters used are different; the methods of correction are different, the styles of presentation are different, and mathematical backgrounds are somewhat different. Moreover, Indian astronomy was influenced much more by the needs of astronomy. Ujjain, an important center of Indian astronomical activities, was taken as having zero longitude.

Astronomy is a practical subject and relies on successive approximation to attain truth. The Hindus displayed a willingness to abandon methods yielding incorrect result and were committed to enhancing them. Starting with Panch Siddhantika, each succeeding work gives more accurate methods than the earlier ones.

Indian astronomy ultimately exercised a greater influence on the development of modern western astronomy than on the Greek or Babylonian system because of the influence of Arab astronomical works on western scholars. Of special importance in Arabic astronomy was the *Khandakhadyaka of Brahmagupta*, written in AD 665 and known to the Arabs as the A/-Arkand. Parameters from this work and Indian planetary theories found their way, through Arabic scholars, in particular through al-Khwarizmi, into Western European astronomy. The use of the zero meridian of "Arian" (Ujjain) is the most glaring example of this influence.

12.4 A Brief Description of Ancient Indian Mathematics

i) *Sulva Sutras*: There are seven sulva sutras known by the name of *Bodhayana*, *Apasthamba*, *Katyayanam*, *Manava*, *Maitrayana*, *Varaha*, and *Vishuta*. They belong to the period 800–500 BC. They explain a large number of simple geometrical constructions imparting knowledge of the following theorems:

a) The diagonals of the rectangle divide it into two equal parts.
b) The diagonal of rectangle bisects each other.
c) The perpendicular, drawn from the vertex of an isosceles triangle to its base, bisects it.
d) A rectangle and a parallelogram on the same base and between the same parallels are equal in area.
e) The diagonal of a rhombus bisects each other at right angles
f) Properties of similar rectilinear figures.
g) The theorem known after the name of Pythagoras, i.e., Pythagoras theorem, states that the square of the hypotenuse is equal to the sum of squares of base and perpendicular. Irrational numbers were first used here and the approximation to $\sqrt{2}$ as $1 + 1/3 + 1/3.4 + 1/3.4.34$ was given. Problems in the construction of altars, which required the solution of simultaneous linear equations with integers, were solved and illustrated.

ii) Jain Mathematics belongs to the period from about 500 to 100 BC. It gives elementary formulae for menstruation, approximate values of square roots (e.g. $\sqrt{10}$ correct to 13 places of decimals), approximation of π $\sqrt{10}$, some concepts of infinity, laws of indices, and the formulae for permutations and combination.
iii) The bakshali manuscript, found at the village Bakshali, near Peshawar in AD 1081, is believed to belong to the fourth-century AD; it contains problems on fractions (method of false position), arithmetic and geometric progressions, solution to indeterminate equations, approximation to surds, solution to quadratic equations, and simultaneous linear equations.
iv) Aryabhata (AD 476) gave the algorithm for finding cube roots, sum of squares and cubes of first numbers, approximation of π by 3.1416, mensuration formulae, and tables for sin 0.
v) Brahmagupta (AD 598) solved the indeterminate equation $Nx^2 + 1 = y^2$ in integers, gave the sum of geometrical progression, determined the right-angled triangles and cycle quadrilaterals with rational sides, and gave interpolation formulae and the formulae for the area of a cyclic quadrilateral.
vi) Mahavircharya wrote his famous work, Ganita Sara Sangraha, in AD 850. It contains problems on arithmetic, on quadratic equations, on the

general formulae for cube root, on unit fractions, and on construction of cyclic quadrilaterals of any given area, and results for ellipses.

vii) *Bhaskaracharya* (born in AD 1114) wrote *siddhanta shiromani* at the age of thirty-six, wherein he systematized mathematics known at that time. It gave formulae for surface area of a sphere and its volume, volume of frustums of a pyramid, problems on permutations and combinations, problems on quadratic equations and indeterminate equations. *Bhaskara* also gave some results which show that he could almost have discovered the calculus.

viii) After *Bhaskara*, new mathematics was not searched except in Kerala, where a great deal of mathematical work was done. Ganit Kaumudi by Narayan Pandita of the fourteenth century and the work of Madhava are pioneers. These were followed by *Tantra Sangraha* and *Aryabhatiyam* of Sakara Varman, *Karana Paddhati*, *Digganita*, and *Kariya Kranakara*. In these works, series for $\sin x$, $\cos x$, $\tan^{-1} x$, etc., are given at least two hundred years before they were found in Europe.

ix) The result $\lim_{n \to \infty} (1^P + 2^P + ... + (n-1)^P / n^{P+1}) = (1/P + 1)$ was also given by these authors. *Karana Paddhati* gives the value of π correct to 10 places of decimals, and Sadratanmala gives it correct to 17 places. Nilkantha also clearly stated, in about AD 1500, that the ratio of the circumference to the diameter can never be expressed as the ratio of two integers.

Hindu mathematicians in ancient India were highly influenced by the Greeks at an early stage and by the Chinese at a later stage, but they also contributed a lot to the development and research of mathematics.

There are two aspects to mathematics. One aspect of it pertains to the quantitative interactions among material objects, making it useful in fields like commerce and economics. On the other hand, it makes assumptions and attempts to apply these theorems in a way that makes sense. From this point of view, mathematics can be considered as philosophy.

12.4.1 The decimal system

The first main concepts of mathematics originated from the counting and weighing and involved the invention of numerals within the Hindu–Arabic system and the introduction of zero, which made the concept of place value. At the time of introduction, the numerals and the zero were not rated as being of particular importance. There were other numeric systems in use, and they seemed to be just as valuable as the ones that gave rise to our well-known symbols [2, 3].

The most significant contribution from ancient India to the progression of civilization is the development of the decimal system of numeration, which encompass the invention of zero. This system is characterized by the utilization of nine digits and zero symbol allocation place value to each digit to represent all integer numbers. This was known as the Decimal System.

A well-known archaeologist, Dr. S. R. Rao, has professed a new theory suggesting that it was the Harappans who introduced to the world the decimal system of measurement and the utilization of gold as a medium of exchange. Dr. Rao is of the view that Harappans produced gold discs that were made in a specific ratio of weight. The unit was known as gunja, corresponding to the Suvarna mashaka mentioned in the Arthasastra. The gold discovered at Lothal and hitherto mistaken for pendants weighed in size precisely 50, 100, 2500, 2750, 2800, 2900, 3000, and 3250 milligrams, thus indicating a decimal system.

Dr. Rao asserted that the Harappans were the pioneers of the decimal system for linear measurement and the weighing of delicate items like gold and precious stone. The 1.7-mm division on the Lothal ivory scale was the smallest known division in the ancient world's decimal system. However, the seals discovered at Harappa, Mohenjo-daro, and Lothal were having distinct marking but were interrelated. Dr. Rao said, "The large building like the Lothal, were proportionately built according to the Lothal linear scale. The measurement system based on the Harrappan possible in to later period. In the Arthasastra, it is stated that the 10 divisions of the Lothal scale correspond to 1 angula." According to his view, Harappans demonstrated impressive precision in measuring the fundamental unit of length and time with the smallest length unit being 1.7 mm and the corresponding mass measuring 50 mg.

There is a school of thought that, despite the lack of direct evidence for extensive mathematical knowledge, the advanced level of the Mohenjo-daro and Harappa civilizations – with their well-planned cities, pucca houses with ventilation, and highly developed irrigation systems – could not have been possible without a corresponding level of mathematical development

12.4.2 Numerals and zero symbols

The profundity of the invention of place numerals is only fully appreciated when it is considered how hard it is to record mathematical advancement using other systems, including the Greek number system I, II, X, C, etc. The Greek approach of representing numbers through geometric segments and the slow growth of mathematics in the West prior to the introduction of Indian system highlight the challenges faced by those who lack familiarity with the Indian decimal system.

The importance of the decimal systems of numeration can best be appreciated in the words of Laplace (1749–1827), one of the greatest mathematicians of all time. He remarked that the idea of expressing all values with nine digits, each carrying both an absolute and a positional value, is so straightforward that its simplicity is precisely why we often don't realize how much admiration it deserves [1, 3].

The symbol for zero was developed after the concept of zero itself had already become widely understood. Therefore, the men who credited the zero symbol and numerals contributed greatly to civilization. The names of these individuals and the time period in which they lived are unknown, as the numerals had been in use for centuries before any records were made. Furthermore, the preservation of mathematicians' names in India only became common at a much later stage in history.

In an article in the American Mathematical Monthly, Professor G. B. Holstead shows that the zero existed in India at the time of Acharya Pingala's work Chhandas-Sutra, a work on prosody before 200 BC. It is impossible to overstate the significance of the zero mark's creation, he says [10, 11]. This points toward an ethereal nothingness–merely a name, a place, or a symbol–but with an inherent, transformative power. Such is the nature of the Hindu tradition, which, wherever it has arisen, has always infused even the most abstract ideas with deep significance. It is comparable to coining nirvana into dynamos.

There has never been a mathematical invention more effective for the overall advancement of power and knowledge. In fact, the zero symbols were used in metric by Pin Gala (before 200 BC) in his *Chhandas-Sutra*.

12.5 Concluding Remarks

(i) There are several differences between Greek and Hindu mathematics. In the first place, the Hindu mathematicians often considered themselves primarily as astronomers, so Hindu mathematics was mainly an extension of astronomical pursuits, whereas Greek mathematicians developed their mathematical concepts independently. The Indian caste structure also contributed to the practically exclusive practice of mathematics among priests and upper caste members. Anyone who wished to study mathematics could do so in ancient Greece [4].

(ii) The achievements of Hindu astronomy are impressive and remarkable, yet they were subservient to some extent to astrology. Western astronomy advanced quickly following the discovery of the telescope. Still, Surya Siddhanta is the basis for the majority of our computations. In view of recent advances in theoretical and observational astronomy, we must revise our antiquated calculation of algorithms [4].

(iii) Indian astronomy did not get as much recognition abroad as it should have, partly because of prejudices of foreigners and partly because of the vagueness of its predictions and lack of sufficient efforts to explain these lucidly.

(iv) Efforts were not made to study the development of mathematics and astronomy in ancient Babylonia, Egypt China, Arabia, etc. to find out the interaction between India and other civilizations.

(v) It may be noted that we have not included 16 sutras and 13 sutras ascribed by Swami Krishna Tirath Maharaj to the Atharvaveda. According to Swami ji from these sutras, the answer to everything can be obtained in one line only. This is because of the fact that there is absolutely no evidence of these sutras being part of the Vedas or having an influence on the later development of mathematics in India or in the world [10, 11].

12.6 Future Scope

As the origin of ancient Indian mathematics is very old and has rich literature for further study, one can explore the application of ancient Indian mathematics in research to connect the graph theory [5, 6] areas and its applications. It can reduce the computational run time of NP-hard problems [7, 9, 12].

Acknowledgments

Gajendera Pratap Singh acknowledges the support from the Department of Science and Technology (DST)-Science and Engineering Research Board (SERB) project (MATRICS ID: MTR/2021/000378), M.Sc. consumable grant, DBT BIC and the Department of Biotechnology, Ministry of Science & Technology, Govt. of India (Project ID: BT/PR40251/BITS/137/11/2021), ICMR Project ID 2021-15450 for providing the research facility. Both authors thank the reviewers and readers for their valuable feedback.

References

[1] Gupta, R.C.: On some mathematical rules from Aryabhatiya, Indian Journal of History of Science, Vol.12 (2), 200-206, 1992.

[2] Hooda, D.S. and J. N. Kapur: Aryabhata Life and Contribtions, New Age Publications, New Delhi, 2001

[3] Smith, D.E.: History of Mathematics, Dover Publications, New Delhi, 1969.

[4] Swarup, A.K. and K. S. Sukla: Proceeding of international astronomical union collogium held in New Delhi, Cambridge University Press, 1987.

[5] Singh, Gajendra Pratap, et al. "Modeling the mechanism pathways of first line drug in Tuberculosis using Petri nets." *International Journal of System Assurance Engineering and Management* 11 (2020): 313-324.

[6] Singh, Gajendra Pratap, et al. "A Graph-Theoretic Analysis on Functional EEG Network in Igraph R." *Decision Analytics Applications in Industry* (2020): 541-555.

[7] Singh, Gajendra Pratap, and Agraj Gupta. "A Petri net analysis to study the effects of diabetes on cardiovascular diseases." *2019 6th International Conference on Computing for Sustainable Global Development (INDIACom)*. IEEE, 2019.

[8] Singh, Gajendra Pratap, and Sujit Kumar Singh. "Petri net recommender system for generating of perfect binary tree." *International Journal of Knowledge and Systems Science (IJKSS)* 10.2 (2019): 1-12.

[9] Singh, Gajendra Pratap, Sangita Kansal, and Mukti Acharya. "Embedding an arbitrary 1-safe Petri net into a boolean Petri net." *International Journal of Computer Applications* 70.6 (2013).

[10] Kansal, Sangita, Gajendra Pratap Singh, and Mukti Acharya. "On Petri nets generating all the binary n-vectors." *Scientiae Mathematicae Japonicae* 71.2 (2010): 209-216.

[11] Singh, Gajendra Pratap, Aparajita Borah, and Sangram Ray. "A review paper on corona product of graphs." *Advances and Applications in Mathematical Sciences* 19.10 (2020): 1047-1054.

[12] Singh, Gajendra Pratap, Sangita Kansal, and Mukti Acharya. "Construction of a crisp Boolean Petri net from a 1-safe Petri net." *International Journal of Computer Applications* 73.17 (2013): 1-4.

13

Identification of Leak Locations With Mathematical Modeling: A Case of Water Distribution Network

Rakesh Chandra Bhadula[1], Chandan Singh Ujarari[1], V. N. Kala[2], and Jaipal Singh[3]

[1]Department of Mathematics, Graphic Era Hill University, India
[2]G.B. Pant Engineering College, India
[3]D.B.S.(PG) College, India
E-mail: Rakeshbhadula.14@gmail.com; ujarari0000@gmail.com; vnkala67@gmail.com; jaipalsingh.dbs@gmail.com

Abstract

The water distribution system can lead to excessive economic loss due to leak conditions in the water distribution system. A mathematical model is formulated to evaluate the leak locations with the help of the equation of continuity and head losses upstream from the hole and head losses due downstream from the hole with leak conditions. The analytical method is used to analyze the leak location in the pipeline system by calculating the pressure variation and velocity of the water in the pipeline system. Pressure head and leak locations are obtained with the help of SCILab software and graphical representation is also done with the help of the SCILab software. With the help of a pressure profile, the leak location can be easily identified after the necessary treatment.

Keywords: Identification of leak, pressure head profile, equation of continuity

13.1 Introduction

The water industry faces many problems due to sudden breakage in the pipeline system. Breakage in pipelines needs to be identified immediately; otherwise some harm can mix in the water due to which many diseases can occur. In such a model, our focus is to obtain the leakage in the distribution system so that no one can be affected by the disease. In this model, n number of leakage points are considered with different flow velocities and different surface areas. Vela et al. [20] stated that various researchers have recently proposed various methods to evaluate the leak location in the water distribution system. Such methods are based on leaks with fictitious valve discharge values and the statistical data that are available in the record. Fontana et al. [7] stated that leakage in the water distribution is still a challenging question for all researchers. van Zyl and Cassa [19] suggested that water leakage can be reduced by the installation of new pipelines or by repairing the pipelines. Bhadula et al. [4] stated that the aging of pipelines also plays a vital role in human life. Bhadula et al. [3] presented a mathematical model, in which it was stated that water quality also depends on the aging of the distribution system. If the distribution system is new, then no leakage can be obtained. Sophocleous et al. [17] stated that the water distribution network contains large geodetic differences and reconstruction of the distribution system can reduce the leakage in the pipeline system, and pressure management can reduce the leakage in the pipeline system. Kammoun et al. [10] conducted a statistical study to identify leak location areas with the help of flow and pressure data from different realistic water demand scenarios. Ainola et al. [1] proposed an algorithm, which was based on a leak between the nodes to pressure in the nodes of the pipe. The authors in [12, 13] analyzed leak location with the help of a transient test based on the time domain or frequency domain. Cabrera and Vela [8] presented a formula for leak deduction in the pipeline network; this formula is expressed as the ratio of leak index and average zone night. Tucciarelli et al. [18] expressed a model in which they investigated water losses per time unit for leakage nodes. Andres and Planells [2] presented leak coefficient, which depends upon the pressure with the total volume of leakage. Pollock and Hsu [15] applied acoustic emission techniques to the evaluation of leakage in the pipeline network. This method was based on the solution of partial differential equations with respect to time. Lay-Ekuakille et al. [11] discussed leakage in the zig-zag distribution system, and this method was based on a diagonalization-method-based algorithm. Roy [16] discussed the leakage in the water pipeline network with hydro informatics; hydro informatics is used with the software-based artificial technology to identify the leakage in the water distribution system. A water hammer equation with leak condition is applied to determine the leak location in the water distribution system. Liu et al. [14] applied Monte Carlo methods to solve the nonlinear pipe model to deduct the leak in the pipeline. Andres and Planells [2] used water hammer equations to deduct leakage. Covas et al. [6] stated that leak deduction has wide applications in the water industry. Jawhar et al. [9] stated that the leak rate is not unique when fluid is passing through the pipe under pressure conditions. In this chapter, the location of leaks is obtained with the help of the equation of continuity and head losses due to the upstream from the holes and losses due to the downstream from the holes. To determine the leak location, *n* leak point is considered with different velocities and different surface areas; then with the help of different pressure losses, leak location is identified.

13.2 Materials and Methods

Here we considered n number of holes (say $n_1, n_2, \ldots, n_n$) present in the pipe section of the pipeline with an inlet and outlet Q_1 and Q_2, respectively. $Q'_1, Q'_2, \ldots, Q'_n$ are taken as discharges obtained from the holes and $A'_1, A'_2, \ldots, A'_n$ are considered as the area of the holes at $n_1, n_2, \ldots, n_n$, respectively. Furthermore, it is assumed that A_1, A_2 are the cross-sectional areas of the inlet and outlet discharges respectively. The equation of continuity for this model can be given as

$$A_1V_1 - (A'_1u_1 + A'_2u_2 + \cdots + A'_nu_n) = A_2V_2 \tag{13.1}$$

$$Q_1 = Q_2 + Q \tag{13.2}$$

where $u_1,\ u_2,\ \ldots, u_n$ are the velocities obtained from the holes and Q is the total discharge obtained from the holes.

Head losses due to fractions can be given as

$$h_f = K_r \frac{LV_1^2}{2gD}, \tag{13.3}$$

where K_r (resistance coefficient) is calculated as $K_r = \frac{16}{Re}$ and $R_e = \frac{\rho VD}{\mu}$, where R_e denotes the Reynolds number and Reynolds number < 4000.

When Reynolds number > 4000, the value of resistance coefficient (K_r) can be calculated from the following formula: $K_r = \frac{0.0791}{Re^{1/4}}$.

In such conditions, the flow in the pipeline will be in a zig-zag pattern, that is, the flow will be turbulent. The head losses due to the upstream from the specific hole can be calculated as

$$H_1 - H_1^{'} = h_f, \tag{13.4}$$

where $H_1^{'}$ is the static head just upstream from the location of the leakage.

Head losses at the length $L - l_i$ due to fractions can be calculated as

$$h_{fi} = K_r \frac{(L - l_i)V_2^2}{2gD} \tag{13.5}$$

$$H_2^{'} - H_2 = h_{fi} \tag{13.6}$$

where $H_2^{'}$ is the static head just downstream from the location of the leakage.

There are *n* holes present in the section of pipe length *L*; then leakage in the *i*th point can be obtained from the following equation:

$$Q_i = F\sqrt{2gE_{i1}}, \tag{13.7}$$

where F is the leak rate, which is the function of internal pressure and $F = kA'_i$, where k is the discharge coefficient.

The overall energy equation can be obtained for the incompressible fluid for the *i*th hole

$$H = Q_1E_{i1} - Q_2E_{i2}. \tag{13.8}$$

Here flow is spouting to the atmosphere, and then the total energy equation can be obtained for the *i*th hole present in the system.

$$H = (Q_1 - Q_2)^3 \frac{1}{2g(F)^2}. \tag{13.9}$$

The value of l_i determines the leak location at the *i*th position of the pipeline system.

$$l_i = \frac{2gD\left(V_2^2(H_1 - H_1^{'}) + V_1^2(-H_2^{'} + H_2)\right)}{K_rV_2^2V_1^2}, \tag{13.10}$$

where $Q_1 = A_1V_1, Q_2 = A_2V_2$.

Eqn (13.10) applies to both turbulent and laminar types of flow problems.

13.3 Result and Discussion

A pipe of length 1000 meters with a diameter of 0.02 meters was taken to analyze the leak location in the distribution system. In this distribution system, the smooth pipe is taken so that the friction coefficient is the function of Reynolds numbers. This model is applicable for both laminar as well turbulent flows in the pipeline systems. Eqn (13.9) shows the relationship between flow and the pressure head. From Figure 13.1, the variation in pressure head can be seen easily, and when the velocity of inlet water is similar to the velocity of outlet water, there is negligible pressure head, but when the difference between the velocity of inlet water and velocity of outlet water is higher, then there exists higher pressure head. This shows that the leakage in the distribution system needs to be identified and must be treated; otherwise some impurities can be mixed in the water pipeline and can be harmful to human life or the pipeline can be blocked due to the leakage in the distribution system. Analysis of pressure head profiles may be a very useful tool to identify the leakage in the water distribution system. If there is no leakage in the distribution system, then the value of the pressure head will be zero, which is justified. Different values of pressure head with different inlet and outlet flows can be seen with the help of Table 13.1.

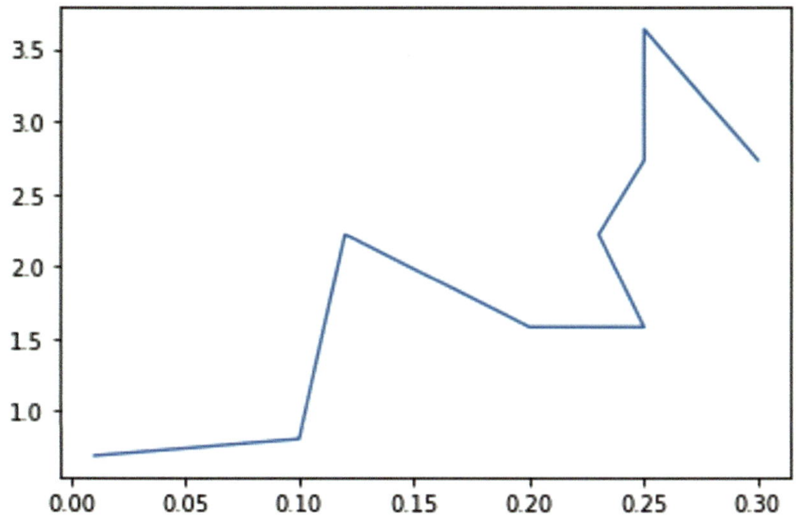

Figure 13.1 Variation in pressure head due to inlet and outlet velocities.

Table 13.1 Variation in pressure head due to inlet and outlet velocities.

S.N.	Inlet velocities	Outlet velocities	Pressure head
1	0.2	0.01	0.69443935
2	0.3	0.1	0.80995988
3	0.4	0.12	2.22252991
4	0.45	0.2	1.58195289
5	0.48	0.23	1.58195289
6	0.5	0.25	1.58195289
7	0.51	0.23	2.22252991
8	0.55	0.25	2.7336146
9	0.58	0.25	3.63844103
10	0.6	0.3	2.7336146

With the help of eqn (13.9), one can understand the leakage due to pressure head profile but cannot identify the leakage in the distribution system. To identify the leakage, the distribution system eqn (13.10) is very helpful. To identify the proper leak location in the distribution system, a 1000-meter pipe was taken with 0.2 meters diameter. When flow is laminar, then leakage in the pipe line can be analyzed with the help of Figure 13.2. From Figure 13.2,

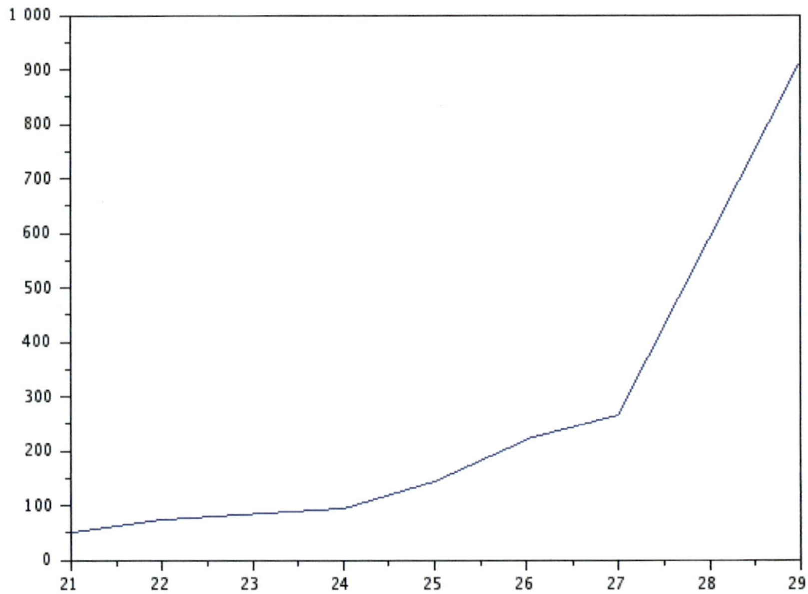

Figure 13.2 Relative pressure at the hole locations.

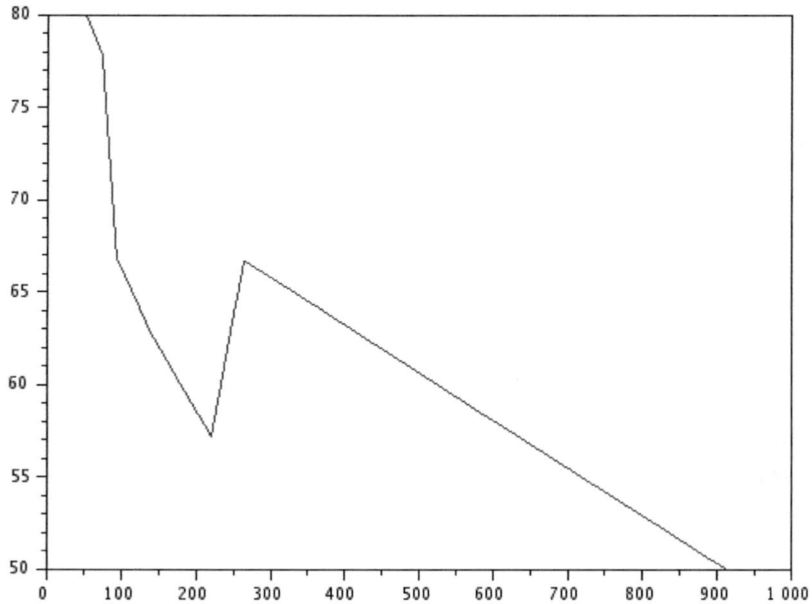

Figure 13.3 Variation between the leak locations and percentage of outlet and inlet flows.

when relative static pressure is 21 meters, then the leak in the distribution system is 74 meters and when the relative static pressure is 29 meters, then leak is 945 meters. It shows that when relative static pressure is increasing, leak locations in the pipe are also increasing.

In Figure 13.3, the percentage of inlet and outlet flows was studied. In such cases, inlet and outlet flows vary between 50% and 80%. When the percentage of outlet and inlet flows was 50%, the leak location was identified and it is 911 meters from the starting point of the pipeline system. When the percentage of outlet and inlet flows is 80%, the leak is obtained 51 meters from the starting point, which is justified

13.4 Conclusion

In this chapter, leak locations were studied with the help of the equation of continuity and head losses upstream and downstream from the hole. The SCILab software was applied to obtain the leak location. Flow profile as well as head loss profile is studied and the relative static pressure as well as the percentage of outlet and inlet flows was studied with the help of SCILab software. It is analyzed that the change in relative static pressure shows a change in the leak location in the distribution system. When the percentage of outlet and inlet flow is higher, then the leak location is obtained near the starting point. With the help of this model, leak locations of the pipeline can be determined, so that necessary treatment can be done, and it is very important in the context of human life safety as well as economic losses.

References

[1] Ainola, L., Koppel, T., &Vassiljev, A. Complex approach to the water network model calibration and the leakage distribution. *WIT Transactions on Ecology and the Environment, 40,* 2000.

[2] Andres, M. & Planells, F. Water losses versus optimal life of a water distribution system's network. *Water Supply Systems, State of the art and future trends, eds. E. Cabrera & F. Martinez, Computational Mechanics Publications: Southampton, UK, Universidad Internacional Menendez y Pelayo: Valencia, pp. 381-409,* 1993.

[3] Bhadula, R. C., Kandari, L. S., Kala, V. N., & Singh, S. J. Chlorine concentration decay in channel due to ageing factor. In *AIP Conference Proceedings* ,Vol. 2481, Issue No. 1, 2022.

[4] Bhadula, R. C., Kala, V. N., Pant, R., Kholiya, D., & Singh, S. J. Ageing of pipe induced chlorine concentration decay: A one dimensional mass transfer modeling. *Materials Today: Proceedings, 46*, 10761-10765, 2021.

[5] Colombo, A. F., Lee, P., & Karney, B. W. A selective literature review of transient-based leak detection methods. *Journal of hydro-environment research*, 2(4), 212-227, 2009.

[6] Covas, D., Ramos, H., & De Almeida, A. B. Standing wave difference method for leak detection in pipeline systems. *Journal of Hydraulic Engineering, 131*(12), 1106-1116,2005.

[7] Fontana, N., Giugni, M., Glielmo, L., Marini, G., & Zollo, R. Real-time control of pressure for leakage reduction in water distribution network:Field experiments. *Journal of Water Resources Planning and Management, 144*(3), 04017096, 2018.

[8] Cabrera, E., & Vela, A. F. (Eds.). *Improving efficiency and reliability in water distribution systems* (Vol. 14). Springer Science & Business Media, 1995.

[9] Jawhar, I., Mohamed, N., & Shuaib, K. A framework for pipeline infrastructure monitoring using wireless sensor networks. *Wireless telecommunications symposium* (pp. 1-7). IEEE, 2007.

[10] Kammoun, M., Kammoun, A., & Abid, M. LSTM-AE-WLDL: Unsupervised LSTM Auto-Encoders for Leak Detection and Location in Water Distribution Networks. *Water Resources Management*, *37*(2), 731-746, 2023.

[11] Lay-Ekuakille, A., Vendramin, G., & Trotta, A. Spectral analysis of leak detection in a zigzag pipeline: A filter diagonalization method-based algorithm application. *Measurement*, *42*(3), 358-367, 2009.

[12] Liou, J. C., & Tian, J. Leak detection—transient flow simulation approaches, 1995.

[13] Liou, C. P. Pipeline leak detection by impulse response extraction, 1998.

[14] Liu, M., Zang, S., & Zhou, D. Fast leak detection and location of gas pipelines based on an adaptive particle filter. *International Journal of Applied Mathematics and Computer Science*, *15*(4), 541-550, 2005.

[15] Pollock, A. A., & Hsu, S. Y. Leak detection using acoustic emission. *Journal of Acoustic Emission*, *1*, 237-243, 1982.

[16] Roy, U. Leak detection in pipe networks using hybrid ANN method. *Water Conservation Science and Engineering*, *2*, 145-152, 2017.

[17] Sophocleous, S., Savić, D., & Kapelan, Z. Leak localization in a real water distribution network based on search-space reduction. *Journal of Water Resources Planning and Management*, *145*(7), 04019024, 2019.

[18] Tucciarelli, T., Criminisi, A., & Termini, D. Leak analysis in pipeline systems by means of optimal valve regulation. *Journal of Hydraulic Engineering*, *125*(3), 277-285, 1999.

[19] van Zyl, J. E., & Cassa, A. M. Modeling elastically deforming leaks in water distribution pipes. *Journal of Hydraulic Engineering*, *140*(2), 182-189, 2014.

[20] Vela, A. F., Espert, V., & Fuertes, V. S. General overview of unaccounted for water in water distribution systems. *Improving efficiency and reliability in water distribution systems*, 61-95, 1995.

Index

About the Editors

Akshay Kumar received his B.Sc. and M.Sc. degrees in science from Chaudhary Charan Singh University, Meerut, India in 2010 and 2012, and his Ph.D. degree major in Mathematics and minor in Computer Science from G. B. Pant University of Agriculture and Technology, Pantnagar, India in 2017. He is currently working as an Assistant Professor in the Department of Mathematics, Graphic Era Hill University, Dehradun in India. He currently teaching LLP and probability and statistics at B.Sc.(H) level and has taught group theory, numerical methods, business statistics, discrete mathematics, engineering mathematics-III, and at PG level has taught OR, linear algebra and discrete structure. He is a regular reviewer for more than 43 international journals including for Elsevier, Springer, Emerald, MDPI, Cambridge University Press, World Scientific, River Publishers, IGI Global and many other publishers. He has published 80 plus research papers in communications in statistics - Simulation and Computation, Emerald, Inderscience, International Journal of Mathematical, Engineering and Management Sciences, IGI Global, MDPI, Cambridge Scientific, Springer Nature, Elsevier Multimedia Tools and Applications and many other national and international journals of repute and has presented his works at national and international conferences. Also, he has published 3 books (edited) with international publishers like River Publishers, De Gruyter and IGI Global. His fields of research are reliability theory, fuzzy reliability, signature reliability and applied mathematics. He was recently awarded the "Outstanding Researcher Award" for 2019-2020 at Graphic Era Hill University Dehradun India and at the 15th and 16th Uttarakhand State Science and Technology Congress 2020-2022 for his significant contribution in academics and research at Graphic Era Deemed to be University, Dehradun, India.

Mangey Ram received his Ph.D. degree major in Mathematics and minor in Computer Science from G. B. Pant University of Agriculture and Technology, Pantnagar, India in 2008. He has been a Faculty Member for around fifteen years and has taught several core courses in pure and applied mathematics at undergraduate, postgraduate, and doctorate levels. He is currently the Research Professor at Graphic Era (Deemed to be University), Dehradun, India. Before joining Graphic Era, he was a Deputy Manager (Probationary Officer) with Syndicate Bank for a short period. He is Editor-in-Chief of International Journal of Mathematical, Engineering and Management Sciences; Journal of Reliability and Statistical Studies; and Journal of Graphic Era University, and Series Editor of six Book Series with Elsevier, CRC Press, Walter De Gruyter, River Publishers and the guest editor and associate editor for various journals. He has published 400 plus publications (journal articles/books/book chapters/conference articles) in IEEE, Taylor & Francis, Springer Nature, Elsevier, Emerald, World Scientific and many other national and international journals and conferences. Also, he has published more than 60 books (authored/edited) with international publishers like Elsevier, Springer Nature, CRC Press, Walter De Gruyter, and River Publishers. His fields of research are reliability theory and applied mathematics. Dr. Ram is a Senior Member of the IEEE, and Senior Life Member of the Operational Research Society of India, the Society for Reliability Engineering, Quality and Operations Management in India, and Indian Society of Industrial and Applied Mathematics. He has been a member of the organizing committee of a number of international and national conferences, seminars, and workshops. He was conferred with the "Young Scientist Award" by the Uttarakhand State Council for Science and Technology, Dehradun, in 2009. He was awarded the "Best Faculty Award" in 2011; the "Research Excellence Award" in 2015; and "Outstanding Researcher Award" in 2018 for his significant contribution to academics and research at Graphic Era Deemed to be University, Dehradun, India. Also, he has received the "Excellence in Research of the Year - 2021 Award" by the Honourable Chief Minister of Uttarakhand State, India, and the "Emerging Mathematician of Uttarakhand" state award by the Director, Uttarakhand Higher Education. Recently, he received the "Distinguished Service Award - 2023" for the subject and nation development by Vijñāna Parishad of India.

For Product Safety Concerns and Information please contact our
EU representative GPSR@taylorandfrancis.com Taylor & Francis
Verlag GmbH, Kaufingerstraße 24, 80331 München, Germany